Praise for *Our Mother Ocean*

"Through overfishing, industrial aquaculture, and poisoning, capitalism is killing ocean life—upon which all of life on Earth depends—but the people who are most directly threatened by this destruction are fighting back, and the rest of us urgently need to join their struggles. That is the story that unfolds in this remarkably detailed but compact book by Mariarosa Dalla Costa and Monica Chilese. To date, awareness of the killing has been mostly limited to the environmental movement. At the same time, awareness of the ways in which capitalism has been slowly destroying traditional communities of those who live by, on, and with the seas has been mostly limited to the peoples of those communities and, in the case of indigenous fishing communities, a few anthropologists. This book not only illuminates the interrelationships between these two patterns of destruction, but also highlights the emergence of a worldwide movement of resistance on the part of some of those most directly threatened."

—Harry Cleaver, author of *Reading Capital Politically*

"In this lyrical text, two political scholars—feminist Mariarosa Dalla Costa and ecologist Monica Chilese—contest the global neoliberal assault on life through the prism of the sea, with its threatened species and courageous fisher peoples. Their impeccable research will bring validation, inspiration, and empowerment to the worldwide struggle of communities for food sovereignty and sustainable, life-affirming cultures."

—Ariel Salleh, University of Sydney, author of *Ecofeminism as Politics* and *Eco-Sufficiency and Global Justice*

"This book about the world's fishermen movement provides us with new insight about a phenomenon that is completely ignored, not only by the mainstream media, but also by independent researchers. Yet the questions it raises about the safeguarding of resources, the right to live, and the satisfaction of needs are strategic. Mariarosa Dalla Costa and Monica Chilese have achieved a great and highly enjoyable book."

—Claudio Albertani, History Department, Universidad Autónoma de la Ciudad de México

"This book is indeed a timely one. With climate change and the exhaustion of natural resources, patriarchal capitalist civilization seems to be coming to an end. The authors remind us that Mother Earth and Mother Ocean are indeed the sources of all life on our planet. Without Earth, no life; without oceans and water, no life. The authors argue that the vital connection between humans and the sea, between humans and the Earth, has been disrupted by capitalist and patriarchal exploitation. The victims of this exploitation, among others, are the small coastal fishermen who lose their livelihood. However, the authors do not stop at analyzing their problems, but show how people everywhere are fighting against this destruction. I warmly recommend this book to all who are concerned about the future of life on this planet."

—María Mies, author of *Patriarchy and Accumulation on a World Scale* and coauthor of *The Subsistence Perspective* with Veronika Bennholdt-Thomsen

"The emergence of the fisher as part of the movement against neoliberal globalization is beautifully understood in this book. I applaud the authors' passionate portrayal of workers on the sea as an organic part of those of us who wish to protect Nature against the rapacious excesses of capitalism."

—George Katsiaficas, author of *Asia's Unknown Uprisings, Vols. 1* and *2*, and *The Subversion of Politics*

"In *Our Mother Ocean*, Mariarosa Dalla Costa and Monica Chilese sound an eloquent warning about the precarious state of not only the planet's fisheries but of the health of the world's oceans themselves. They foreground the dilemmas facing fishermen's movements in various continents, caught as they are between economic imperatives, the need to fish sustainably, and the pressures of multinational capitalism. This book is a thoughtful and necessary call to action."

—David Gullette, Simmons College, author of *Dreaming Nicaragua*

"There is no apocalyptic randiness in this amazing account of horror. Instead we get a call as rigorous as it is passionate for what we all need to do now. The authors distill for the reader the almost overwhelming documentation they use in their very solid exploration of the subject, covering almost every aspect of it, and they share their insights in an elegant and direct style. The world fishers movement, brilliantly described here—the biggest fishers movement in history— begins to do for Mother Ocean what Via Campesina, the biggest farmers movement in history, is doing for Mother Earth. In resisting the new enclosures, hundreds of millions of people are thus attempting to stop the devastating activity of corporate capital, in order to sustain their ways of life, and ours. They need both our awareness and our action. This is the book we need for both."

—Gustavo Esteva, author of *Grassroots Post-Modernism*

"*Our Mother Ocean* is an engaging and critical effort by Mariarosa Dalla Costa and Monica Chilese, who bring attention to the concerns, questions, and struggles relating to the seas and their remarkable social, economic, cultural, and ecological importance to human beings. This appealing book not only questions our relation with the sea but aims to raise consciousness about the way we live our lives and the ecologic problematic we all face globally. It is in this context that the authors brilliantly relate the path of the movement of fishermen, a movement born in the seventies in India that has now spread all over the world. Under the banner of food sovereignty, this movement fights the neoliberal predatory assault and view of sea life as mere products, while struggling to establish a different relationship with the sea, a sustainable relationship with this source of life that ensures the protection of both the small coastal communities who 'have always lived on the sea and of the sea' and the sheltering of the beauties, habitats, and ecosystems of our mother ocean."

—Massimo Modonesi, professor of history, sociology, and Latin American studies, director of the Department of Sociology of Universidad Nacional Autónoma de México, Mexico City

OUR MOTHER OCEAN

Our Mother Ocean: Enclosure, Commons, and the Global Fishermen's Movement

Translated by Silvia Federici

Originally published in Italy as *Nostra madre Oceano: Questioni d lotte del movimento dei pescatori* in 2005 by Derive Approdi, Rome.

ISBN: 978-1-942173-00-7
LCCN: 2014949635

10 9 8 7 6 5 4 3 2 1

Common Notions
131 8th St. #4
Brooklyn, NY 11215
www.commonnotions.org

Cover design by Josh MacPhee/Antumbra Design
Interior design and typesetting by Morgan Buck/Antumbra Design
www.antumbradesign.org

Printed in the United States of America.

OUR MOTHER OCEAN

Enclosure, Commons, and the Global Fishermen's Movement

Mariarosa Dalla Costa and Monica Chilese
Translated by Silvia Federici

CONTENTS

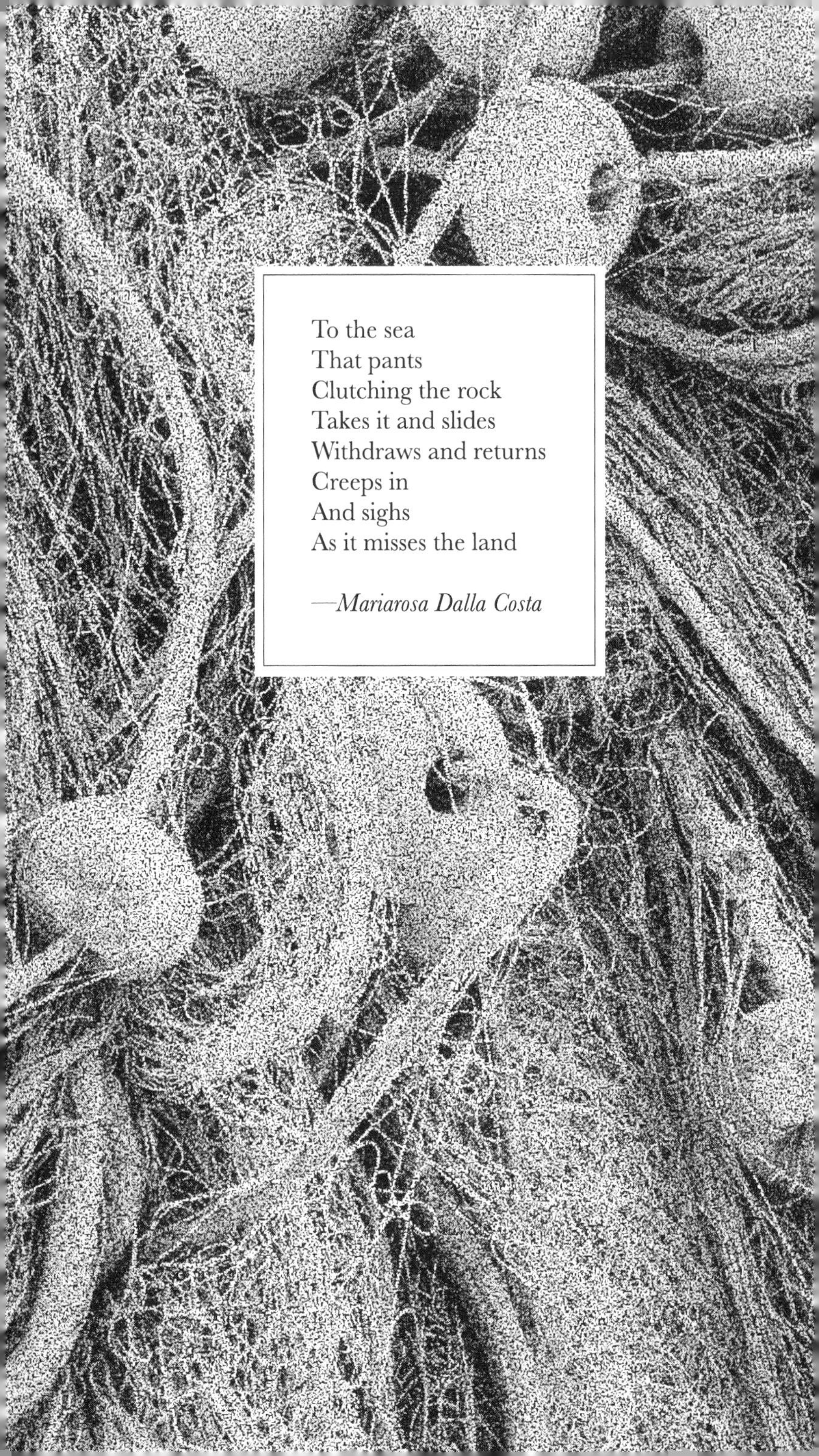

To the sea
That pants
Clutching the rock
Takes it and slides
Withdraws and returns
Creeps in
And sighs
As it misses the land

—Mariarosa Dalla Costa

TRANSLATOR'S PREFACE

Silvia Federici

Our Mother Ocean, the fruit of a collaboration between internationally renowned feminist political theorist Mariarosa Dalla Costa and sociologist Monica Chilese, addresses one of the most crucial issues of our time: the ongoing destruction of our seas. Since time immemorial, the sea has served as the source of our life on the planet, as provider of not only livelihoods but knowledge, beauty, spiritual strength. All of this is now at risk, however, of dying, as the oceans are turned into the poisoned receptacle of the world's waste. Of this destruction *Our Mother Ocean* traces the main aspects, providing a great wealth of information about the consequences of industrial fishing, aquafarming, marine pollution, and the continuing failure of the institutional initiatives presumably predisposed to protect the Earth's ecosystems. What makes the book special, however, is that its denunciations of the many ways in which the ocean is depleted of its immense wealth is accompanied by a passionate reflection on what the sea has signified in the history of humanity, as reflected in its literature, its myths, its philosophies and religions. The book also provides a history of the rise of the first worldwide fishermen's movement, reminding us that the protection of Earth's waters is as crucial in its economic, political, and spiritual implications as that of its lands and forests.

What the ocean signifies for life on the planet is powerfully evoked by Chilese's initial chapter, which takes us through the changes in the relationship between human beings and the sea, from its representation as a symbol of the infinite, the unknown, the sacred, a "theater of the universal struggle between life and death," to its utilitarian reduction to a "usable object," a container of exploitable resources to be freely appropriated and commercialized. Chilese's central theme is that the ocean is far more than "a mass of water." It is the producer of the oxygen we breathe, of the clouds that cool the planet, of the food on which most of the world's populations depend for their survival. Against this background, her description (in Chapters Two and Three) of the dangers now threatening the life of the sea should give a jolt even to readers cognizant of the ecological devastation produced by industrial technology, shaped as it is by competition and the quest for the maximization of profit.

It is now commonly acknowledged that our seas are being emptied of their fauna and flora due to overfishing, that the relentless pouring of industrial and urban contaminants in their waters is destroying the coral barriers and creating miles-long archipelagos of trash, and that the list of ecological catastrophes affecting the oceans is expanding by the day. Barely had we recovered from the horrors generated by the BP oil spill in the Gulf of Mexico when the Fukushima disaster, whose end is not in sight, confronted us with the nightmarish prospect of thousands of tons of radioactive material being daily poured into the ocean's body, undoubtedly causing a further collapse of its network of living organisms. Still, Chilese's detailed documentation of the devastation caused to fish stocks and the marine environment by the operations of trawlers, aquaculture, the mining of the seabeds, and the release of all sorts of contaminants into the seas paints an alarming picture, precluding any complacency or the hope that the present degradation of the ocean may be reversed through minor reforms.

What we learn from her account is that in fishing, as in other spheres of life, technological progress has only expanded the capacity for destruction. Modern fishing boats now use sonar, satellites, ecosounding gadgets—all technologies developed for military purposes—to ensure that nothing escapes their nets, although much of the fish thus caught are ultimately rejected and thrown back, now as waste, into the water, for only what fetches a good price on the market is considered worthy of being retained. Another important lesson offered by Chilese is that the same dangers that threaten the life of the ocean also threaten the survival of the communities whose livelihoods depend on the sea, causing the disappearance of knowledge, forms of employment, and communal relations. This last theme, however, is most developed in the second part of the book, where Mariarosa Dalla Costa examines the effects of the industrialization of fishing on coastal communities, especially in the Global South, and their resistance to this process, leading to the formation of the world's first fishermen's movement.

Dalla Costa is a writer widely known in U.S. radical and academic circles since at least the early seventies, when her foundational essay "Women and the Subversion of the Community" inaugurated a feminist critique of Marxism that transformed not only the debates within the women's liberation movement but Marxist theory itself, also providing the manifesto for the International Wages for Housework Campaign of which Dalla Costa was, with Selma James, the cofounder. Based on the redefinition of women's unpaid domestic labor as a quintessential form of capitalist production, insofar as it is a production of labor power, Dalla Costa's theoretical work has since placed the question of social reproduction at the center of her political and theoretical activism. However, while in the 1970s the focus of her analysis was the condition of the "housewife" in her relations to capital and to the male waged worker, in later years her work has been mostly concerned with the manifold forms of resistance that women and men across the planet are engaged in, in opposition to the neoliberal assault on their means of subsistence and reproductive systems.

In a variety of articles published between the late eighties and the present, she has returned again and again to this theme, increasingly inspired in her perspective by the work of ecofeminist writers and the struggles of indigenous people, starting with the Zapatista movement. At the core of her theoretical engagement in these contexts have been the devastating impacts of the World Bank's structural adjustment programs on the lives of people across the continents and the central roles of the struggles over land and food sovereignty in the construction of new commons, developing her thinking about them hand in hand with her participation in the antiglobalization movements. *Our Mother Ocean* is a logical step in the urgent and ongoing discussions we must have about a subsistence perspective, social reproduction on a planetary level, and the commons, for only by artificial distinctions can we separate the planet's lands, air, and waters.

Dalla Costa's contribution to the book, a long essay dedicated to the rise of the fishermen movement, focuses on the social consequences of the "blue revolution"—from India, Pakistan, Bangladesh, and Sri Lanka to Malaysia, Indonesia, the Philippines, Canada, the United States, Mexico, Honduras, Japan, Norway, and Madagascar. Coastal communities everywhere are forced to abandon their lands, as they cannot compete with corporate fleets, or for instance, they must make space for the cultivation of shrimp. Indeed, paraphrasing Thomas Moore's famous comment on the English enclosures, we can say that "shrimp" today are "eating human beings," in a bizarre twist in which the need to satisfy the palate of those who can pay exacts costly human sacrifices, such as the loss of land and enslavement to the brutal work regimes that the shelling and processing of the popular fish demand.[1] Echoing similar denunciations by Vandana Shiva, Dalla Costa contrasts the irreparable damages produced by the new fishing technologies with the creativity of the traditional methods they replace, which for centuries guaranteed the livelihood of millions of people. What she describes is a war against farming and fishing communities that relies on collusion between the fish/aquaculture industry and the local political elites and police forces that kill and torture at their service, all to bring shrimp and other chosen fish to the tables of people thousands of miles away.

Since the late eighties, however, resistance has begun to mount on the sea as on the land, and it is one of the merits of this book to have brought to the foreground this struggle and, as a new international political subject, the world fishers' movement first created in India in 1997 that since then has spread to every part of the world. Dalla Costa's recognition of the importance of this movement, generally ignored by most "global justice" theoreticians and activists, is timely. Not only is the expropriation of the marine wealth of the populations of Africa, Asia, and Latin America by multinational corporations proceeding at an accelerating pace; in the "North" as well we are witnessing a sustained assault on the remaining fishing commons, ironically conducted under the guise of conservation and protection against overfishing.

Exemplary in this regard is the attack that in recent years has been waged on the fishermen and fishing communities of New England by the Regional Fishery Management Councils (of the National Oceanic and Atmospheric Administration) that, in arguing for the need to rebuild fish stocks, have introduced "catch-sharing" programs, i.e., fishing quotas that have been established on the basis of previous fishing history that privileges large boats and makes it impossible for small owners to survive. If this succeeds, the New England fishing industry will undergo a "historic change," marking the end (in the words of a Maine fisherman's journal) of "a tradition of fishing rights older than this nation."[2] Thus, at the moment, New England fishermen are fighting desperately—but quite alone—against this de facto privatization of the fisheries in an attempt to save their source of livelihood and the existence of their communities. Taken by itself, their struggle may seem hopeless. But this is where a book like *Our Mother Ocean* becomes most important. For it broadens our political horizon and helps us realize that resistance against the enclosure and destruction of the ocean has now become a global phenomenon.

NOTES

1. "[Y]our sheep that were wont to be so meek and tame, and so small eaters, now, as I hear say, be become so great devourers and so wild, that they eat up, and swallow down the very men themselves. They consume, destroy, and devour whole fields, houses, and cities." From Sir Thomas More, *The 'Utopia' and the History of Edward V*, Maurice Adams, ed. (London: Walter Scott, 1980).

2. *Fishermen's Voice*(December2009),availableat:http://www.fishermen voice.com

INTRODUCTION

Mariarosa Dalla Costa

The fishermen's movement that emerged in the seventies in the south of India has today a planetary dimension and in its record a heroic history of women and men. But in many countries in the Global North, it has not had a proper resonance. Yet it represents a great story of commitment, sacrifice, and poetry rich with universal meaning because of the questions it raises concerning the relations between work and the safeguarding of resources, the satisfaction of needs and the awareness of limits. They are questions of love and respect but paid with the blood shed in ever harsher clashes between the right to live and profit obtained at the price of a death sentence for many populations. At stake is the respect for life, not only that of other human beings but that of other living beings, which is manifest in poetic terms in the documents that mark the path of the movement.

This book, born of the same passion for the sea and the desire to share it with others, is intended to make a contribution to this path. We wish, first of all, to stress the polyvalence of the vital functions that the oceans represent. For oceans represent not only food but medicines, raw materials, climate, environment, biodiversity, and culture. We also wish to draw attention to the main changes in the problematic relation that has marked the history of human beings with the sea: their gradual approach to it until the recent "conquest" of the marine depths and the irrational exploitation of the riches of the abysses, leading to the depletion of this great reservoir of nature, which is now impoverished and altered. We aim to raise consciousness about the questions involved in the relation of human beings with this opulent source of nutrition and life, established first through the activity of fishing, from the casting of the fishing line for sport to commercial trawling. The book is intended to spread knowledge about a certain problematic among social circles not already directly engaged. Indeed, a remarkable awareness has grown about the questions of agriculture and land, and alternatives have been developed—thanks mainly to the initiatives taken by social subjects from the South of the world who have come to the North to denounce the consequences of productivism and technologism for real agricultural productivity and their lives. Yet similar issues relating to the sea and fishing have remained more hidden, more enclosed inside a discussion between the fish workers and those who oversee this sector.

Twenty years after the first Conference on Sustainable Development in Rio (the United Nations conference on environment and development that led to the adoption of Agenda 21, a blueprint for sustainable politics to be implemented in the twenty-first century), all the main issues remain practically unresolved. The condition of the ecosystems has worsened instead of improved, and economic inequities and social injustices have increased. Chapter 17 of the agenda, which concerns the protection of all the seas and coastal areas as well as international regulation on this matter, must gain a new authority to effectively protect the marine and adjacent coastal ecosystems and the populations who live there. Article 17.3, which calls for the protection of the exclusive economic zones (the sea up to two hundred miles from the coast) for the benefit of the area's residents, is far from being implemented. Destructive activities of various types—not only large-scale industrial fishing, characterized by gigantic catches and bottom trawling that ruins the seabeds—continue, constantly gaining new areas of predation. As the northern seas are impoverished, the prow has turned to the seas of the south. And as the catches of many seas on the planet have been diminished or depleted, it is the coastal areas of the developing countries that are now devastated by industrial installations for the farming of shrimp and other fish that destroy the environment and with it the nutritional resources of the inhabitants of the regions. These installations increase the pressure on marine life, because a large amount of fish feed is needed for these farms, and this continues to be caught with industrial boats and bottom trawlers. At the Summit on Sustainable Development held in Johannesburg in August 2002, an agreement was reached that foresaw the recovery of fish stocks by 2015 and the abolition of dangerous subsidies, but the text appears weak with regard to the plan of action, and according to environmental organizations it represents a step back with respect to the promises made in Rio in 1992 and Chapter 17 of Agenda 21. Despite the coming into effect of the Convention on the Law of the Sea of November 1994, despite the Code of Conduct of the Food and Agriculture Organization (FAO) for Responsible Fisheries with the relative International Plans of Action of 1995, and the convention on the conservation of highly migratory fish stocks of 2000, responsible fishing is still far from being achieved, while excessive fishing, outlawed fishing practices, and industrial aquaculture continue to compromise "the blue world" and its fragile coastal ecosystems.

What is ever more in danger is the whole fish patrimony, the great commonwealth that for thousands of years has guaranteed a livelihood to human settlements with modest economies but that are rich in their harmony with nature, rich therefore in all the goods that this harmony provides. More and more frequent are ecological catastrophes that, given the industrial and touristic alteration of landscapes and ecosystems, find all defensive walls destroyed and in the same places add the victims of catastrophes to the victims of impoverishment. The tsunamis in Southeast Asia in 2004 and 2005 are only some of the many tragic and dramatic examples of this in our time. The causes of poverty in the world are in any case quite evident. In addition to the choices made by the agricultural policies imposed on the countries of the South, we want to draw attention to those that characterize the politics of fishing. Precisely in this context, then, it is extremely significant that a great movement of fishermen has emerged as an important but largely unknown protagonist in a still ongoing movement of movements in the Global South. The new subjects of this movement place at the center of their activity the safeguarding of the organic connection between the craft of fishing and the maintenance of the ecosystem, with its great variety of marine, marshy, riverine, animal, and vegetable species that give meaning and provide a rich relationship between ecology and work on the sea. The fishermen's movement moves under the banner of food sovereignty. This is a population's right to produce its own food and to maintain access to the natural sources that generate it through the cooperation between human beings and nature. From this follows the right to maintain those small economies that, being based

upon a friendly relationship with the ecosystem, preserve and utilize its resources within limits allowing for their renewal by continuing to practice crafts that are the fruit of that knowledge. This is the case for not only the coastal communities of India, to which we give special attention in this text, but also for various communities of fishermen in the advanced countries, which now develop their movements' analysis alongside those of the former, finding common solutions and paths. The main objective of Via Campesina, the largest network of farmers who carry across the planet the flag of food sovereignty, is to guarantee the right of populations to have access to basic resources, first of all the land and the common goods it contains, from natural seeds to the water of lakes and rivers. In complete harmony with these goals, the fishermen's movement demands first of all access to the seas, the lakes, and the rivers, to be able, by means of these resources, to continue those activities that, managed in a traditional manner, have guaranteed life for millennia.

Fishermen want to protect their economies, they want to be able to improve them and not be compelled to alter them or be expelled by economic choices that do not recognize their rights to continue to live. But the right to have access to the land and the sea is devoid of meaning if what remains is only a parched land or a sea without fish. The common good, then, requires that initiatives be taken to reconstitute it. This, in my view, is one of the most significant questions emerging in the path of the fishermen's movement. If the farmers, in rebellion against the industrial-biotechnological model of agriculture, give time to the soil to regenerate itself so that it can host an organic agriculture, similarly fishermen often take initiatives to reconstitute the ecosystem of a given region to allow for the repopulation of that site by the species that characterized it, and which make it possible to reanimate the devastated small economies of the region. This is the case with the people in the Philippines who have gone to study the techniques of reforestation and have replanted the mangroves, which cradle many species of fish, wherever they had been cut down to make room for aquaculture tubs that were later moved to a different place. These decisions signify above all the determination not to accept as inevitable and irreversible the new, higher level of development that is imposed, in this case industrial aquaculture, with its devastation of species and human beings, and to restore instead the type of development previously adopted. This is a creative, friendly development, in contrast to the destructive one that is now prevailing, given that the promised creation through destruction of the latter has been demonstrated not to be credible. Their activities echo the words of the late Thomas Kocherry, who states in a speech he delivered upon receipt of the Sophie Prize in Oslo in 1999: "From Canada to Senegal, Brazil, Pakistan, South Africa, the fishermen of the North and South, all victims of globalization, are trying a new development paradigm . . . where the capacities and technologies of the natives are valorized, where small is accepted as beautiful and sustainable, and simplicity becomes a way of life, paying due respect to indigenous cultures."

At the 1999 WTO Ministerial Conference in Seattle, demonstrators who were dressed as turtles and whales put on trial the predatory approach of large-scale fishing while denouncing the nonobservance of the sentence of the Supreme Court in India that since 1996 has ordered the closing of all the existing industrial installations along the country's coast. However, the phenomenon of intensive aquaculture has for quite some time been devastating numerous coasts of tropical countries. In 2003 in Italy, for the first time in recorded history, the usual herds of tuna did not arrive in front of the few remaining traps of Favignana and Bonagia in Sicily, while a new phenomenon developed: the caging of red tuna, which is "fattened up" in large cages offshore to be ready to replenish the Japanese markets. The sequence is then repeated: the destruction of the cycles allowing for a spontaneous reproduction of life, penury instead of abundance, and the strategic creation of a fictitious abundance now replicated in the new marine prisons as well as in the coastal ones.

In the presence of these trends there is a great urgency for new forms of intervention. The social and political consequences of this conquering approach have been first of all the loss of food and livelihood. But there has also been the loss of habitat, beauty, and awareness. Some of the paths that need to be taken have already been indicated by environmental movements and experts in this sector: the abolition of perverse subsidies in the fishing industry—that is, the abolition of those forms of aid that end up being used to increase overfishing; the adoption of provisions to eliminate all types of pollution; the creation of protected marine areas even on the high seas; the introduction of measures promoting sustainable fishing; and the elimination of threats to ecosystems. On one side, therefore, it is important to call for a change in politics; on the other it is necessary to contribute to the formation of a different consciousness among women and men. While the discourse on the quality of material life and the meaning of work—central for the institution of a different relation to the land—is reemerging in political debate, we also have to restore a different relation with the sea.

The ocean is immense and deep. To find our mother again we cannot challenge it but must approach it with love and respect, crossing the blue planet on the same boats as the small communities who have always lived on the sea and of the sea.

ONE

"THE FOUNT AND SCOURGE OF OCEAN LIFE"

Monica Chilese

"MAN" AND THE SEA: FEAR, ADMIRATION, AND EXPLOITATION

Over the course of centuries poets and writers have narrated the beauties of the seas, the secrets of their depths, and the evolution of the particular relation that ties human beings to the marine environment. Caffarena's reflections on this relation have been a fundamental point of reference for my own rethinking of the views held about the sea by famous authors in history.[1] From the "throbbing expanse, symbol indeed essence of the unknowable"[2] to the "vehicle of a prodigious life,"[3] the sea has gradually been accepted by humans. Its ambiguity, in fact, as a source of life on one hand and on the other a domain of death makes people perceive it as at once a malevolent and propitious power. The perception of it as something hostile and in itself terrifying, caused above all by its boundless immensity and ungovernability, has populated the imagination of ancient mariners and those who have, by means of myths, monsters, legends, and tales, turned it into a theater for "the universal duel between life and death."[4]

Traveling through history we find that, as early as Babylonian mythology, Tiamat, the queen of the abyss and the ocean, constantly threatens to destroy all creation. In the epic of Gilgamesh,[5] the hero in search of immortality is warned against crossing the sea: "This ocean is the ocean of death and no mortal has ever navigated it."[6] The Bible's representation of the sea in the classic era immensely contributes to its negative image: the Book of Genesis,[7] with its tale of creation and the deluge; the Book of Job,[8] the Psalms, and the Book of Isaiah,[9] with the Leviathan, the sea monster symbol of the devastating force of evil, have a profound effect

on the collective imagination. The ocean is the threatening primordial sea that is opposed to the cosmic order; it is "the kingdom of the unfinished, vibrating and vague extension of chaos."[10] For Plato the ideal city must rise quite far from the sea; he considered it "the enemy of good civic laws."[11] According to Horace, who abhors the *dissociabilis* ocean that divides men, navigation is a challenge to the divinity, and its tempestuous sea, lusting for the shipwrecked, is a theater for bloody clashes.[12]

"Some cartographers dare to localize, on theologically inspired maps of the world, almost infernal places populated by dragons, true kingdoms of sin."[13] The monsters of the depths of the sea, creatures that terrorize navigators, represent the fantastic embodiment of the brute force of the sea, with its tempests, seaquakes, fogs, and flat calms, and become the manifestations of the omnipotence of its elements. As Victor Hugo wrote in 1869, "The water-spout is the monster. . . . [H]ere it comes"; a monster is also the tiger wave, "a ferocious, final wave that arrives at the proper moment, creeps for a while as if stretching with its belly over the sea, then leaps, roars, shrieks, falls on the endangered ship and dismembers it."[14] The ancient shore then becomes the receptacle for marine excrement; on the shore the sea purges itself, vomits its monsters; amber, for instance, was considered the most precious and extraordinary residue of marine excretions until the seventeenth century. Other much-celebrated symbols of the dangers that the ancient mariners could meet during their navigations are the mermaids and Scylla and Charybdis. The myth of the mermaids points to the danger posed by dead calms that in the summer season may immobilize ships, causing the phenomenon of the mirage: "[N]ever the ocean is more savage than when it resembles a pool."[15] They embody also, for adventurers of the waters, the charm of the unknown that, however, can lead the inexperienced to death.

Scylla and Charybdis,[16] instead, represent the risk of navigating in straits, namely that of Messina, and along rocky coasts in tempestuous days, when currents and vortexes can push a ship against the rocks. As for the monster Charybdis, even today the area is famous for its squids and polyps and jellyfish, which are found there in great quantity. Against this hostile universe, which hosts appearances of the devil, true rites of exorcism are practiced. From the Adriatic coast to the banks of Terranova in Calabria, a whole set of magical practices, propitiatory actions, and rites have been elaborated, of which today we still have the traces: to consult an astrologer; to get onboard a witch doctor armed with a knife to cut the hurricane while pronouncing magical formulas; hiding coins under the base of the mast. . . .

In the case of the Venice lagoon, two superstitions are worthy of being remembered: the exorcisms against the stings of poisonous fish[17] and those against waterspouts. After being stung the wounded goes to fetch an exorcist, an old fisherman who knows how to mark. After making the sign of the cross he pronounces the mysterious words: "I mark you and may God heal you." And after the rite: "How do you feel, my flesh?" "Better, I believe, better my old man, I had pains so strong that I gnawed the saint's image. May God reward you, my old man!" "Have faith, my son, have faith in God! Not a leaf of a tree moves if God does not want!"[18]

Thus, the practical aspects connected to navigation and the not remote danger of possible attacks by pirates and corsairs again sharpened the terror of the defenseless sailors. Life onboard the ships was characterized, for a long time, by conditions of extreme harshness: hardships; food privations; the lack of hygiene that, at times, had fatal consequences. Here is what the Vicentine Antonio Pigafetta narrated in his travel logs:

> Wednesday, November 28, 1520, we came out of this strait, plunging into the Pacific sea. We stayed three months and twenty days without taking any refreshment. We ate biscuits that were no longer biscuits but their dust, with handfuls of worms, because they had eaten the good part (it smelled strongly of the urine of mice), and we drank yellow water

> already putrefied for many days, and we ate certain ox skins that were above the main mast, so that the mast would not break the stay, they were very hard because of the sun, rain and wind. We would leave them for three or four days in the sea, and then we would put them a bit on the coals and we ate them this way, many other times [we ate] sawdust. We sold the mice, each for half a ducat, if only we had some! But beside all other hardships, the worst was this: some [sailors] had their gums growing over their teeth, above as well as underneath, so that they could not eat and died. Nineteen men died of this illness and the giant together with an Indio of the land of the Verzin. Twenty-five or thirty men became ill, some in their arms, others in their legs or other places, so that few remained healthy. Thank God, I did not develop any infirmity.[19]

The bishop António de Guevara, in his *Arte del marear* (Pamplona, 1579), gave the following advice to those who had to face the waves: "On the galleys you settle as you can, not as you like it. You will not find benches to stretch upon, nor chairs to sit on. Do not dream of leaving your shoes and your socks, do not go without your mantle, it is the only mattress on which you can count. Passengers and sailors lay down at random and the head of one can be found next to the feet of another."[20]

Only a few years before, the French poet Eustache Deschamps had written:

> When the tempest is chasing us
> It's necessary to go under the deck
> We feel grabbed by the throat
> And we vomit for the stench.[21]

It is precisely under the deck that you risk the worse diseases, typhus and scurvy *in primis*,[22] so that a simple wound could lead to death. "The gray winter ocean, lugubrious and cold, sums up every kind of fear, it feeds the terror of being caught by an unforeseeable death, without the comfort of the last sacraments, far from the circle of one's family, and being consigned, body and soul, without a burial, to the infinite waves that never rest."[23] To die at sea then, with an "unwept burial," represents another obsessive source of anguish that feeds new legends, in which the cry of the sea gulls is interpreted, for instance, as the cry of the souls that rest on the seabeds.

The repulsion for this world of the abyss, and the fear of the fury of the sea, continue until the thirteenth through fourteenth centuries, and though they do not disappear, they are slowly transformed, thanks to the greater safety of navigation and the increased knowledge of the marine environment, into an admiration that exalts nature as prodigal of gifts. The golden age of discoveries, under the impulse of Henry the Navigator,[24] provides a different perspective, bringing people closer to the real sea, far from the imaginary involucre in which it had been enclosed. One goes in search of faraway lands and unexplored islands, and in a few decades maritime expansion opens Europe's eyes. The representations of the oceans and its coasts are enriched by new experiences; the insular myth grows with its dreams of happiness to the point that "language is not enough to tell and the hand to write all the wonders of the sea," as Columbus is alleged to have written during his 1492 voyages. In his *Utopia*, Thomas More, to illustrate his hypothesis of a perfect society, gives voice to a phantom character, a travel companion of Vespucci. Itlodeo, his creation, describes a dream that arrives from faraway seas, a paradisiacal island, an ideal model of civilization. The biblical sea monster of infinite power becomes, a century later, in Hobbes's *Leviathan* (1651), the symbol of the state, a sort of "mortal god" to whom all people owe the utmost reverence on this Earth.

Later, natural theology[25] further contributed to helping erase the negative images of the biblical interpretation. It is said that the ocean and its shores have been predisposed by God for the well-being of "man," and in them we admire the power of the Creator. The very composition of the water reveals the Creator's benevolence. The salt protects it from pollution, ensuring the survival of the fish and the health of the shore; thanks to its properties it is possible to preserve food; it hampers frost, favoring fishing and the reproduction of marine creatures. Tides exist to clean up the shore and to facilitate navigation. In this perspective, even tempests acquire a positive value; their turbulence serves to improve the air, to purify it, renew it. Navigation is exalted. It brings men together; it contributes to the work of missionaries; it expands knowledge on Earth and facilitates trade. We are now in the presence of a receptacle of wonders consisting of an infinity of resources but more and more pauperized by the predatory work of men.

Beginning in the middle of the nineteenth century, thanks to that sort of "spatial revolution"[26] that attributes to the sea a new dimension—depth—begins what today has become an intense and irrational exploitation of its treasures. The sea from then on is no longer approached with respect, and even less is it considered the sacred house of the divine. It becomes an object of study and manipulation, reified for the satisfaction of human beings and their predatory instincts typical of modernity. "To every progress of technology corresponds an advance in the ferocious barbarity of extermination."[27] Techno-scientific knowledge and more functional boats increase the possibility of seizing resources that, already compromised by the 1930s,[28] have become today, as we will see in the following chapters, more and more scarce.

A MASS OF WATER, A RESERVOIR OF WONDERS

The essential element of the system that makes life possible,[29] however, finds its own life, its living surfaces, and its populous abysses compromised and threatened. Indeed, the sea, with its extension equal to 71 percent of the planet Earth,[30] cannot be considered only a mass of water. The gradual approach of man to this primordial element, characterized by enormous risks and often nefarious consequences—as documented by the history of the main civilizations of the past that developed along the coasts—confirms the enormous importance of the sea for the perpetuation of our species. Let us then synthetically illustrate its great potentialities, which often have made it a theater of political conflicts and have led to an increasing appropriation of its resources.

In the first place, the main part of atmospheric oxygen is produced by unicellular microscopic organisms that float over the surface of the oceans. It is the photosynthetic plankton,[31] a complex and heterogeneous mixture of organisms, that are the first step in all the marine alimentary chains and contribute, together with the photosynthesis performed by plants on the mainland, to providing us with precious gas. The oceans, moreover, function as great basins for the containment of dissolved gases that contribute to regulating the composition of the air we breathe. As Aldo Morrone informs us, "marine algae produce, in variable quantity, a gas called dimethyl sulfide *(DMS)* that reacts with the oxygen present in the air over the oceans, forming minuscule solid particles. These aerosols of phosphate provide a surface on which the water vapor can condense to form clouds. And clouds keep the planet cool, making the solar radiation bounce off into space."[32] Oceanic currents, fundamental for the climate, transport enormous masses of water across vast distances, providing a constant interchange between the equatorial waters and the cold polar ones. Without the warm Gulf Stream, for example, the temperate zones of northwestern Europe would resemble the subarctic zone. It is enough to think of the meteorological phenomenon of El Niño, which is produced by a modification of the oceanic currents, and of its devastating effects, to understand how their alteration can cause true disasters.

Oceanic and coastal ecosystems demonstrate, moreover, the generosity of nature that manifests itself here in all its domains: animal, vegetable, mineral. The animal domain, which is divided into thirty-seven great groups, almost all marine, is largely represented by reptiles, mammals, mollusks, crustaceans, and an enormous variety of fish. To this day, there are more than "twenty-five thousand known species of bony fish and new ones are constantly discovered."[33] We must underline that this wealth is present above all in tropical waters. The Pacific and Indian oceans host many more species of fish and mollusks than the Atlantic.

Marine fauna represents undoubtedly a great reserve of food: fish, in fact, satisfies about 16 percent of the world's protein requirement,[34] with peaks of 30 percent in Asia. At least one billion people draw from the sea the main resources for their nutrition."[35] "Over two hundred million people depend on fishing for their livelihoods, [and] over 75 percent of the fish consumed by people comes from the harvest of wild species in natural ecosystems."[36] Fish is the main source of the omega-3 polyunsaturated fatty acids, which is a widely demonstrated anticarcinogenic agent. These fatty acids contained in fish also ensure cellular elasticity, preventing cardiovascular diseases.[37] The alimentary chain on which we rely, whose equilibrium we unfortunately modify, represents therefore an immense patrimony of biodiversity and can be synthetically described in this way: Among the primary producers we find the fitoplankton that proliferate in vast spaces of the ocean, forming true "fields"; the first-grade consumers are represented by the zooplanktonic organisms; among the secondary ones we can place the carnivorous and mistivorous fish; then come the predators, the tertiary consumers such as tuna, cod, sharks, squid, marine birds, seals, and others as well. Organisms can live in the pelagic areas, free to move inside the mass of water, or carry on their existences on the bottom, in the benthic zone. Here reigns the most complete darkness; there are no more producers and nourishment is scarce, so benthic organisms must draw most of their sustenance from the rain of debris that comes from the surface. Not all abyssal alimentary chains rely, however, on the sedimentation of debris. Innumerable invertebrates and small fish feed themselves by rising to the surface in the night, then descend to deeper waters before the daylight. These migrants are chased and larger animals prey on them.

Since it is difficult to find nourishment in deep waters, the animals of the abyss have developed curious adaptations to feed themselves. The bacteria found in the lower strata are fundamental, because in consuming the organic debris that falls to the bottom, they free a large quantity of mineral substances, thus enriching the waters of the lower strata.[38] They can be divided into phytobenthos, which consists of algae, macrophytes zoosteracea, and bacteria, and zoobenthos, which is composed of only animal organisms. The latter constitute the epifauna that is divided into: sessile groups of animals that are constantly fixed on the bottom; sedentary ones, with little capacity for movement; and vigile ones, having the possibility of true motion. In the abyss, in any case, the species that can survive are few and diversity diminishes rapidly. However, thanks to discoveries made in the 1970s, we have found that the fauna present in the vicinity of volcanoes active on the seabed is more widespread than what was once thought, a true treasure as far as the variety of living organisms is concerned.[39] The following figure illustrates the trophic chain of the ocean, helping us to better understand the relation between producers and consumers that is typical of every marine ecosystem.

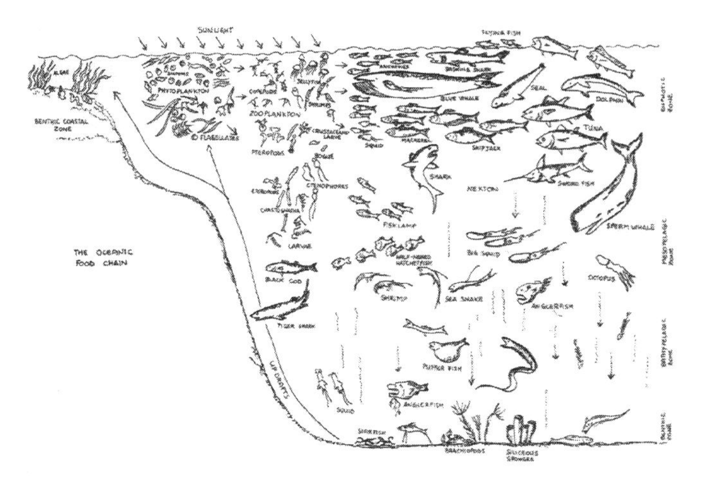

This scheme represents the alimentary chain of the ocean. The main part of the organic substance on which the life of the ocean is based is synthesized in the superficial strata (called "euphotic zone") and illuminated by many varieties of phytoplankton. These microscopic vegetal cells are the food of the herbivorous zooplankton and of some small fish that are in turn the prey of the nekton, which is composed of active predators. The rain of organic debris (represented by the arrows and points) and vertical migration provide the main source of food for the various inhabitants of the mesopelagic, the bathypelagic, and benthic zones. The benthic coastal zone also benefits from some large fissile algae and the drainage of the coast. The typical ascent of the waters along the continental embankment (see the long arrows on the left) furnish to the phytoplankton the organic substances that have been decomposed by bacteria on the bottom of the sea.

Source: J. D. Isaacs, "*Forme di Vita nell'Oceano*," *Le Scienze*, n. 16 (December 1969): 72–73.

To the ocean ecosystem, swarming with life, we must add the very fertile coastal ecosystems: brackish wetlands, marshes of mangroves, estuaries, and coral barriers. Their waters, very rich in light and nutritional elements, provide the precious plankton that is fundamental for most of our fishing activities. The abundant and manifold vegetation, present in the brackish wetlands of the temperate zones and in the mangrove marshes of the tropics, provides for the nutrition and the development of that marine alimentary chain that we have already analyzed, and also protects the coasts from erosion. As Myers informs us (1990), the vegetables of the rocky coasts give us, moreover, alginate composites that contribute to the composition of hundreds of finished products, such as plastics, waxes, deodorants, soaps, detergents, cosmetics, coloring substances, varnishes, lubricants, alimentary stabilizers, and emulsifiers. Mangroves, which cover more than half the coastal tropical zones, in addition to protecting them from hurricanes are a true cradle of marine life, offering shelter for sea turtles, dugongs, manatees, crustaceans, and many fish. Coral barriers formed by small polyps, capable of producing a sort of rigid scaffold of calcite, are also nurseries hosting the quintessence of tropical biodiversity, with a variety of life similar to that of the Amazonian forests. Some of their components, moreover, are used even in the treatment of some diseases.[40] The Wallace Line, in the heart of the Indonesian archipelago, offers, for instance, an incomparable variety of submarine forms of life. Four hundred different species of corals and almost one fourth of the world fish species provide 60 percent of the animal proteins in the diet of the Indonesians who benefit from these wonders. After having closely observed them, the English naturalist Alfred Wallace, who discovered them, wrote:

> The clearness of the water afforded me one of the most astonishing and beautiful sights I have ever beheld. The bottom was absolutely hidden by a continuous series of corals, sponges, actiniae, and other marine productions, of magnificent dimensions, varied forms, and brilliant colors. . . . The bottom was very uneven, rocks and chasms, and little hills and valleys, offering a variety of stations for the growth of these animal forests. In and out among them moved numbers of blue and red and yellow fish, spotted and banded and striped in the most striking manner, while great orange or rosy transparent medusae floated along near the surface. It was a sight to gaze at for hours, and no description can do justice to its surpassing beauty and interest. For once, reality exceeded the most glowing accounts I had ever read of the wonders of a coral sea. There is perhaps no spot in

the world richer in marine productions, corals, shells and fish, than the harbor of Ambonya.[41]

Leaving Wallace to his marvelous species, one cannot forget, as far as coastal ecosystems go, the water mirrors of the lagoon of Venice, a city "that seems born from out of nothing, between sky and water, for it is not the reasonable mixture of land, light, water, and green that geography regularly offers us in all the cities of the vast world."[42] The lagoon, which extends from the left bank of Port Fossone (Adige) to the source of the Isonzo, hosts rich populations of fish of marine origin and from brackish waters and is the favored site for the reproduction of many species, some even of high economic value, such as gilthead, bass, plaice, and mullets. Its temperature, rather high in the summer because of the sunlight the lagoon receives, its shallow depths, and its notable supply of nutrients, which come from fresh water, make it one of the most productive environments among all aquatic ecosystems. Thus, from the time of the first settlements, its inhabitants have considered it an indispensable element for their survival and their economic and political development.

Coastal zones, in addition to their natural processes, are also important because they have always provided favorable conditions for the development of anthropic activities. The southern coasts of the North Sea have been inhabited since 2000 B.C. We still find there traces dating back to the Neolithic time. Ancient and enormous shells of mollusks have been found along the lagoon coasts of Scandinavia and Britain, and from the Gulf of Guascogne to the river Tago, an almost certain sign that they were used as food; similar evidence can also be found in India, North Africa, America, and Australia.[43]

To this journey, that we could continue stressing ever more the beauty and richness of such patrimony, we must add that, as far as the mineral kingdom is concerned, submarine deposits provide almost one quarter of all combustible gas and petroleum. Polymetallic nodules and sulfurs, first discovered in 1958 by the Downwind Expedition on the Blake Plateau off the coast of Florida, contain, moreover, important metals for the realization of steel alloys: managanese, nickel, copper, cobalt, molybdenum, and vanadium. Before the discovery of nodules in Thailand and Indonesia, tin had already been extracted in shallow waters.[44]

Thanks to the properties of marine products, it was estimated in 2000, the value of marine resources was equivalent to $21 billion a year, 70 percent more than the value of the resources drawn from the earthly ecosystems.[45] However, for this patrimony of humanity, as mentioned earlier, *mala tempora currunt* [TN: 'bad times are upon us."]

THE CRISIS AND ENCLOSURE OF THE *ARCHÉ* OF LIFE

After a long period during which human beings lived in equilibrium with nature, capable of using its fruits without impoverishing it and without altering its habitats, we are now suffering from the illusion that man is the master of creation and not its guardian. Our domination of nature means that the cosmos is reduced to a usable object, to the service of a cultural animal that takes the liberty to ignore its rules.[46]

The transition from traditional societies to industrial ones is sketched schematically in the following.

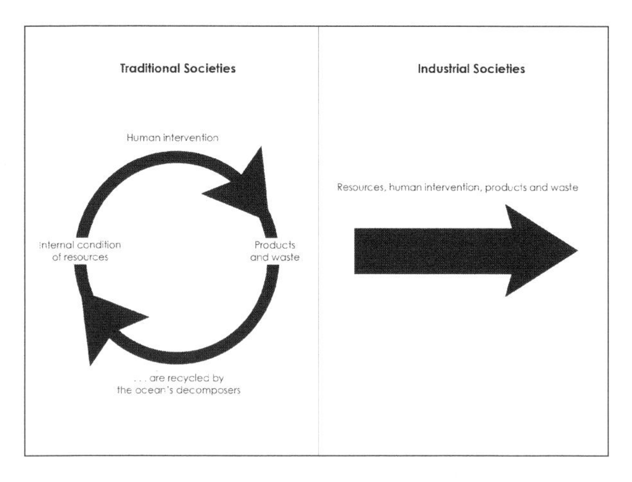

Source: C. Fontana-M. Giacci, *Gli alberi e la foresta*, vol. A (Padova: Cedam, 2001), 84.

This schema illustrates how men behave as predators. Natural resources are used, exploited without any concern for their renewal. The mentality associated with industrialization considers nature an inexhaustible source of wealth to consume, in conformity with short-term economic calculations. Scientific reductionism, which attributes value to only our species, considers nature as something inert to be dominated[47] and has led to the impoverishment of that luxuriant, vivid, manifold life that characterizes the planet. Behaving like the so-called fugitive species—lemmings, ploceidae, grasshoppers—that exploit the resources of a given territory until its exhaustion and then leave, every year we are responsible for the loss of thirty thousand species of plants and animals.[48] The estimates on the Red List of Threatened Species of the IUCN (International Union for the Conservation of Nature)[49] confirm this responsibility, giving evidence for the existing threat to diversity. With our shortsighted opportunism and the irrationality of our behavior we have turned upside down essential ecological processes, so that ecosystems are on the verge of collapse. By sacking and sacking we have negated the relationship between human beings and nature, impoverishing our very identity and alienating it from its environmental context. We are, therefore, a fugitive species, growing at the dizzying pace of 77 million people per year[50] who, by now, no longer know where to flee. This situation undermines not only ecological but also social stability. The erosion of biodiversity, caused by the destruction of habitat and an economic and technological disposition that has imposed uniformity and lucrative monocultures, has led to the drying up of our knowledge, threatening the principles on which the production and maintenance of life are based: diversity, symbiosis, reciprocity.[51] Blinded by profit, we have gone so far as to lose our capacity to distinguish one animal or plant from another, forgetting how species interact with each other and with the environment,[52] and worse yet, we have neglected our ties with the animal world.

In this context, blue spaces, their depths included, as we have explained, acquire above all a capitalist value in which the most profitable species are appropriated without any scruples for the dominant alimentary system.

Giuseppe Notarbartolo, the president of the Central Institute for Research and Technology Applied to the Sea (Istituto Centrale per la Ricerca Scientifica e Tecnologica Applicata al Mare), has declared in an interview that the health conditions of the oceans are not at all good, due to our actions, and we must "hurry up" to safeguard them.[53] The more extensively capitalism develops, the more quickly the conditions of the oceans deteriorate. In 1997 it was estimated that 50 percent of the world's population lived in coastal regions; by 2025 that figure may easily double. Meanwhile, more than 70 percent of the world's megalopolises are located along coasts.[54] Urban economic activities thus (mis)managed and their waste burden, growing at the rhythm of 300 million to 500 million tons per year,[55] inevitably threaten maritime and coastal areas. The nonrenewable use of animal and vegetable species, which are being exploited at a quicker pace than their capacity to repopulate themselves, and the chemical and toxic pollution within the trophic chain, and the sound pollution that even causes lesions on cetaceans,[56] represent the highest form of unsustainability, and to this day, they are problems that have not been resolved. We must not forget here the effects of global climate change that have, for example, caused a temperate sea such as the Mediterranean to become populated with tropical species at the expense of indigenous ones.[57]

We see, then, that Baudelaire's lines, dedicated to the best proof of the sublime—"free man, you will always love the sea / it is your mirror: in it you contemplate … your soul"[58]—strongly contrast with the threats that place its intimate, long-preserved riches at risk. They now appear empty, rendered meaningless by the intensity with which capital commands a contradictory human activity that praises but at the same time destroys.

NOTES

1. A. Caffarena, *Governare le onde. La prospettiva della cooperazione internazionale* (Milano: Franco Angeli, 1998).
2. A. Corbin, *L'Invenzione del mare* (Venezia: Marsilio, 1990), 12.
3. J. Verne, *Twenty Thousand Leagues Under the Sea* (Blackburg, VA: Wilder Publications, 2008).
4. This is an expression by French historian Jules Michelet: ". . . beautiful spectacle, great, fascinating. The universal duel of Love and Death that takes place on the earth is nothing if compared to what occurs on the seabed. Down there, where it has an unthinkable greatness, it terrifies with its fury, but if we observe it closely we discover a great harmony and a surprising equilibrium. This fury is necessary. In this exchange of substance, so quick and stupefying, in this prodigality of death, resides salvation." (quoted in V. Serra, ed., *Le parole del mare. Libero viaggio nell'oceano letterario* (Milano: Baldini & Castoldi, 2002), 109.
5. Gilgamesh (spelled also Gilgamesc or Gilgames) is the legendary king of Uruk. He is the greatest Mesopotamian hero, the protagonist of the Assyrian-Babylonian poem that carries his name. (G. Ciavorella, *Mito, Poesia e Storia* (Torino: Il Capitello, 1990], 626).
6. Ibid., 15.
7. Genesis 1:1–31; 6:11–23.
8. Job 3:8; 38:16; 40:25.
9. Isaiah 27:1.
10. A. Corbin, *L'Invenzione.*
11. P. Janni, *Il mare degli Antichi* (Bari: Edizioni Dedalo, 1996), 17.
12. A. Corbin, *L'Invenzione.*
13. M. Mollat du Jourdin, *L'Europa e il mare* (Bari: Laterza, 1993), 75.
14. V. Hugo, *L'uomo che ride* (Milano: Garzanti, 1988), 103–4.
15. Ibid., 119.
16. In Greek mythology Charybdis is the name of the daughter of Poseidon and Gaia who, guilty of having stolen from Heracles the oxen of Geryon, was transformed by Zeus into a sea monster that in the form of a deadly whirlpool swallowed navigators. It is located in front of the rock of Scylla, on the extreme edge of the Calabria coast.

    So with much lamenting we rowed on and into the strait; this side lay Scylla; that side, in hideous fashion, fiendish Charybdis sucked the salt water in. . . . A shy terror seized on the crew. We had looked her way with the fear of death upon us; and at that moment Scylla snatched up from inside my ship the six of my crew who were the strongest of arms and sturdiest. When I turned back my gaze to the ship in search of my companions, I saw only their feet and hands as they were lifted up, they were calling to me their hearts' anguish, crying out my name for the last time.

Homer, *The Odyssey*, XII, trans. by Walter Shewring 222–224; 229–235 (Oxford: Oxford University Press, 1980), 148.

17. Some fish elaborate poisonous substances through their organisms, secreted by special glands and communicated at the exterior through aculei, thorns and stings, wounding those who handle them. The most feared by fishermen are: the *matana*, the *murenes*, the *varagnola*, the scorpion fish. The *varagnola* and the scorpion fish are those fish that frequently sting the fishermen of Veneto, as they are very common in the High and Middle Adriatic Sea. The *varagnola* is feared above all because, with thorny rays that stem from its dorsal spine and its operculum, it can wound seriously with the poison it puts into the lesions. Once it was believed that these fish were the souls of those who had died in mortal sin who, wandering through the spaces without peace and being deprived of the heaven's delights, cause damage and ruin to the workers of the sea. (Administration of the Province of Venezia, ed., *La pesca nella laguna di Venezia* (Venezia: Albrizzi, 1981).
18. Ibid., 102.
19. Baldi, G., et al., *Dal testo alla storia. Dalla storia al testo*. vol. II, 1, (Torino: Paravia, 1995), 375.
20. S. Bono, Corsari nel Mediterraneo (Milano: Mondadori, 1993), 115.
21. du Jourdin, *L'Europa*, 288.
22. This disease, caused by a lack of vitamin C, struck sailors in particular because there were never enough fresh vegetables loaded on the boats. In the eighteenth century scurvy still decimated the crews of the English navy, so much so that it put the power of the nation in danger. In 1757 a doctor from Portsmouth, James Lind, intuited that the disease could be caused by a lack of fresh food in the alimentation of sailors and recommended that a daily portion of fresh citron be added to the food supplies of the ships. At that time, however, it was difficult to imagine that such a simple remedy could suffice for such a devastating disease. We had to wait, then, until at the beginning of the nineteenth century, when all English ships were ordered to keep a supply of citrons onboard to finally defeat scurvy (H. Ruesch, *Imperatrice nuda* [Milano: Rizzoli, 1976]).
23. A. Corbin, *L'Invenzione*, 21.
24. Henry the Navigator, heir to the Portuguese throne, son of John I (Oporto 1394–Sagres 1460). He transformed his residence, Villa do Infant in Tercena Naval, into an arsenal and nautical institute, from which he directed his captains' navigation and African colonization.
25. "Between 1690 and 1730, in the West a cultural movement

spread that, from the 17th century on, will be called in France 'natural theology,' and in England 'psycho-theology.'" This system of thought was based on the belief in mysterious correspondences between the physical and the spiritual worlds, the human and the divine, between man, the microcosm, and the universe, the macrocosm. (A. Corbin, *L'Invenzione*, 41)

26. "What is a spatial revolution? Man has a specific consciousness of its space, subject to great historical transformations. To the manifold forms of life correspond equally varied spaces. . . . Life and the world manifest themselves not only in a different light but also in other dimensions, depths and horizons." (C. Schmitt, *Terra e mare* [Milano: Giuffré, 1986], 55). The spatial revolution modifies the collective perception of the sea; it allows us to embrace with our imagination its depths and inaugurates a new era in the relations between "man" and the sea. The birth of modern oceanography, which simultaneously realizes this revolution, enables the scholar for the first time to go beyond the surfaces of the seas and study the mass of water and its content. According to Schmitt, moreover, the factor that most contributed to developing an awareness of this new dimension of the sea was the appearance of submarines during World War I (A. Caffarena, *Governare le onde. La prospettiva della cooperazione internazionale* [Milano: Franco Angeli, 1998], 30; C. Schmitt, *Il nomos della terra* [Milano: Adelphi, 1991]).

27. J. Michelet, *Il mare* (Genova: Il Melangolo, 1992), 22.

28. A. Vallenga, *Ecumene Oceano* (Milano: Mursia, 1995), 92.

29. The majority of scientists agree that life itself originated more than three billion years ago in the oceans, with the evolution of simple bacteria and unicellular algae. This microflora received from the sunlight and the nutrients contained in the water what was necessary to generate the complex molecules of living tissues. The atmosphere of the Earth, rich in water vapor, methane, azote, and carbon dioxide, did not filter the ultraviolet rays of the sun, and only the depths of the sea, thanks to the filtering action of the water, could provide the conditions for the development of living beings. They were at first anaerobic bacteria, capable of living in the absence of oxygen, as it occurs in the fermentation processes, and in a second moment (about a billion years after), after the appearance of the first unicellular algae, living beings capable of transforming carbon dioxide and water into sugar by using the energy of the sun. These producing beings, from which will descend the evolutionary line of plants, generated oxygen, profoundly modifying the atmosphere. Oxygen, an ecological catastrophe for anaerobic bacteria, created bit by bit the conditions for life on Earth. The thin stretch of atmosphere that surrounds the Earth's globe was constituted over billions of years and is the essential condition for life. (C. Fontana, M. Giacci, *Gli alberi e la foresta*, vol. A [Padova: Cedam, 2001], 10).

30. The Pacific Ocean by itself covers 36 percent of the planet, the Atlantic Ocean, 21 percent, and the Indian Ocean, 14 percent. We know moreover that the ratio of water to land is higher in the southern hemisphere than in the northern (N. Meyers ed., *Atlante di Gaia. Un pianeta da salvare* [Bologna: Zanichelli, 1989]; *The Gaia Atlas of Planet Management* [London: Gaia Books Ltd., 1985]).

31. "The [word] plankton derives from the Greek plankton and it indicates the particles and the organisms that live suspended, floating on the water. . . . A first subdivision can be made between vegetal and animal organisms respectively called fitoplankton and zooplankton. The fitoplankton live on the surface and in relation to the transparency of the water, never descending more than 200 meters. It is composed of microscopic algae (Bacillariophyceae, Dinophyceae, and Crysophyceae), often unicellular, that utilize the mineral salts dissolved in the water by means of photosynthesis. It multiplies very rapidly, through simple cellular division, and the velocity of production increases remarkably when the conditions of illumination and nutrition are favorable. . . . Zooplankton is formed by a great variety of forms of marine life, protozoa Foraminifera and Radiolaria, coelenterates, crustaceans, mollusks, chetognatha that feed on fitoplankton. . . . With regard to the dimensions of the organisms it can be further subdivided into: macroplankton, mesoplankton, microplankton, nanoplankton, and picoplankton. With regard to the variations in the light and temperature, we can speak of epiplankton that lives in the superficial strata, and bathiplankton that is exclusively composed of animal organisms and populates the strata beneath 200 meters, where the absence of light prevents the growth of plants. In www.galileits.it/ipertesti/grazing/plandes.htm

32. A. Morrone, *L'altra faccia di Gaia. Salute, migrazione e ambiente tra Nord e Sud del pianeta* (Roma: Armando, 1999), 257.

33. N. Eldredge, *La vita in bilico. Il pianeta terra sull'orlo dell'estinzione* (Torino: Einaudi, 2000), 118.

34. B. Paltrinieri. *Una risorsa da rispettare*, August 31, 2000, in www.enel.it/enel/magazine/boiler/boiler05/html/articoli/Paltrinieri-apertura.asp.

35. *Proteggere l'integritá degli oceani e delle coste*, in www.wwf.it/ambiente/sostenibilitá/pianoazione_7.asp.

36. V. Shiva, *Stolen Harvest: The Hijacking of the Global Food Supply* (Cambridge, MA: South End Press, 2000), 37.

37. *I numeri della pesca nell'Italia e nel mondo*, July 18, 2002. In www.wwf.it/news1872002_3744.asp. For the treatment of cardiovascular diseases in antiquity they used cardium, a bivalve mollusk commonly called "heart." The medical properties of mollusks do not belong only to folk beliefs. Some studies conducted in the 1970s have scientifically proven that one can extract from a species of bivalve mollusk, called *Mercenaria mercenaria*, a substance

that inhibits the development of cancer in laboratory mice. Some substances contained in gasteropods are effective instead against some species of microorganisms that are resistant to penicillin. (S. Giannatasio, *Le conchiglie terapeutiche*, in www.mareinitaly.it/curiosita_v.php?id=56).

38. *La catena alimentare*, s.d., in www.galileits.it/ipertesti/grazing/catene.htm.

39. N. Eldredge, *La vita in bilico.*

40. *La scomparsa dei coralli*, February 20, 2003, available at: www.wwf.it/news/2022003_3602.asp; *Barriere preziose per la biodiversitá*, February 22, 2003, available at www.wwf.it./news/2222003_1220.asp.

41. The area takes the name from Alfred Wallace (1823–1913), the naturalist who explored it in 1869. The document where this text appears, which is reported on later in the text, is found in an article by John C. Ryan, an environmentalist and journalist and former Worldwatch Institute researcher: "Indonesia's Coral Reefs" (April 15, 2001), which is available online at: www.worlwatch.org/node/502.

42. F. Braudel, *Il Mediterraneo*, vol. 1 of *La Méditerranée et le Monde Méditerranéen a l'époque de Philippe II,* 3 vols. (Milano: Bompiani, 1987), 244. Originally appeared in 1949; revised several times.

43. P. Lasserre, "Coastal lagoons. Sanctuary, Ecosystem, Cradles of Culture, Targets of Economic Growth," in *Nature and Resources* 15 (1979).

44. A. Caffarena, *Governare le onde*, 33

45. B. Paltrinieri, *Una risorsa da rispettare*, August 31, 2000, available at:www.enel.it/enel/magazine/boiler/boiler05/html/articoli/Paltrinieri-apertura.asp.

46. In this regard, Mainardi has underlined a very interesting distinction between biological and cultural evolution. They do not have the same temporal structure; one is slow and conservative, the other rapid and innovative, placing the first at a notable disadvantage. "The remote cause of every equilibrium, of every extinction is the impact of the cultural irrational behavior of our species on a world of livings that evolve biologically. We are a destabilizing species precisely because in a world of beings, animal, vegetable, that evolve biologically, we are the only great specialist producers of culture. . . . The biological times necessary for the evolution of the natural adaptive counterstrategies are too slow to adapt to the rapidity of change produced by man." (D. Mainardi, *L' animale irrazionale. L'uomo, la natura, e i limiti della ragione*, [Milano: Mondadori, 20001], 126–27).

47. C. Merchant, *The Death of Nature: Women, Ecology and the Scientific Revolution* (New York: HarperCollins Publishers, 1980).

48. N. Eldredge, *La vita in bilico.*

49. The International Union for the Conservation of Nature (IUCN)

has published the following estimates. There are 11,167 species threatened with extinction, at the time of writing 121 more than with respect to the year 2000: they are composed of 1,137 mammals (24 percent of the total), 1,192 birds (12 percent), 742 fish (7 percent of the fish species), 293 reptiles (4 percent), and 157 amphibians (3 percent); 1,932 species of invertebrates are also included. There has been an increase also in the number of plants that are threatened. Presently there are a total of 5,714 (there were 5,611 in 2000): 1,046 are in critical danger, 1,291 are in danger, 3,377 are vulnerable. In the case of Italy, the list includes 183 species of animals and vegetables, 96 at risk: 14 mammals, 9 fish, 6 mollusks, 5 birds, 4 reptiles, 4 amphibians and 41 invertebrates. *More than 11,000 species at risk in the world*, October 16, 2002, available at: www.wwf.it/news/16102002_276.asp.

50. From the report of the Worldwatch Institute, *State of the World 2002*, available at: www.wwwf.it./news/2132002_9945.asp.
51. V. Shiva, *Monocultures of the Mind: Perspectives on Biodiversity and Biotechnology* (London: Zed Books, 1993).
52. V. Shiva, *Biopiracy: The Plunder of Nature and Knowledge* (Cambridge, MA: South End Press, 1999).
53. Cited in C. Giammatteo, *Bisogna fare in fretta*, May 3,2001, available at:www.enel.it/il/magazine/boiler/boiler39/html/articoli/Giammatteo-Notarbartolo.asp.
54. M. Carbone, "*Le milieu marin et le développement durable*," *Le Courrier ACP-UE*, n. 193 (Juillet–Aôut 2002).
55. *Oltre 11,000 specie a rischio nel mondo*, Ottobre 16, 2002, available at: www.wwf.it/news/16102002_276.asp.
56. In July 2002 a group of fifty-eight whales were found stranded on a beach in Western Australia. Despite the help of rangers and volunteers who intervened, fifty-four died, and scientists believe that naval instruments may have disoriented the cetaceans, bringing them to the shore. Available at: www.corriere.it/Primo_Piano/Scienze_e_Tecnologie/2002/07_luglio/27/balene.shtml.
57. We do not know, however, to what measure climate change, pollution, and overexploitation contribute to the spreading of these species. The first migrated fish was found in 1902 in Hàifa (Israel) not far from the Suez Canal, which had been opened thirty-three years earlier. Since that time, coming mostly through this opening, about fifty-three species of "alien" fish have arrived in the Mare Nostrum out of a total of 250 vegetable and animal organisms. They arrive also from the Strait of Gibraltar, attracted by the warm waters, or attached to the keel of ships, or through the ballast waters released into tWhe sea. Some of the nonindigenous species are: the puffer fish that arrives from the Pacific Ocean through the Red Sea; the multicolored scar fish (or parrot fish); the scorpion fish typical of the Red Sea; the leaf fish that comes from the Suez Canal; and the barracuda that is widespread above all in the Atlantic and the Pacific.

The Central Institute for Scientific and Technological Research Applied to the Sea (ICRAM) that observed the phenomenon of "tropicalization" for the first time in 1995 is presently monitoring it to measure the impact of alien species on the ecosystems affected. (B. Paltrinieri, *Mediterraneo un futuro da tropico*, August 31, 2000), available at: www.enel.it/enel/magazine/boiler/boiler05/html/articoli/Paltrinieri-ambientimarini.asp.

58. C. Baudelaire, *I fiori del male* (Milano: Feltrinelli, 2003), 57.

TWO

THE IMPOVERISHMENT OF THE MARINE FAUNA: SOCIAL AND POLITICAL PROBLEMATICS

Monica Chilese

LET US TRY TO ANALYZE, then, the vulnerability of the sea, considering in particular the impoverishment of the marine fauna that endangers its future existence. Indeed, while once we could say *fiat iustitia, pereat mundus* (let there be justice, the world can die)—in which *mundus* naturally meant a *quid*, the renewable essential element of a whole that is never ruined—today we cannot say this any longer, because the possibility of destruction has become too real.[1]

We have explained by what steps we have arrived at this situation; now we deepen our analysis of the dangers that threaten "the vast reservoir of nature,"[2] namely: excessive fishing; illegal fishing; the environmental impact of industrial aquaculture; and pollution. These are all activities that reduce the resources of the sea to the bare minimum, undermining its preservation.

EXCESSIVE FISHING

Fishing represents one of the main economic activities in many countries in both the North and the South. The objective, when the fishing industry is managed in a rational way, is to reach an optimal level of production while guaranteeing the renewal of the marine biological resources. Since 1950, thanks to the refinement of fishing techniques and the possibility of working and freezing the fish caught on the boats, the global catch of fish for human sustenance and the production of animal feed has witnessed a constant increase.[3] Not even the maximum quotas established in the 1970s have been respected.

Technological changes, moreover, are in continuous development, due above all to the subsidies that states give to the industry. These subsidies, which should create jobs in poor coastal areas and favor the development of the fishing industry, in most cases are devoted to creating new technology that leads to overfishing.

According to estimates of the World Bank, these subsidies amount to a total of $20 billion a year.[4] In addition, the fishing fleets for deep-sea fishing, with their large fishing boats, have exported industrial fishing to developing countries, endangering the future of local fishing communities.

To this growing technological advance we must add the problem of free access to the majority of fishing waters. Between 1957 and 1982 the part of the sea placed under national jurisdictions, with consequences for its exploitation, has passed from three miles of the territorial sea (a nautical mile corresponds to 1,852 meters) to two hundred miles of the exclusive economic zone (EEZ).[5]

Coastal states—insular or archipelagic—exercise, therefore, total control over the surface of the sea, over the watery mass, the seabed, the subocean, and all the resources there contained. However, the EEZ, a prescription by the United Nations Convention on the Law of the Sea that was supposed to guarantee rational exploitation under the responsibility of the coastal states, has turned out to be not very effective from the viewpoint of environmental protection.[6] In the chart following we can well identify the main fishing areas.

After years of industrial expansion, the commercial fishing industry has stabilized, because harvesting has reached a point that exploitation exceeds the reproductive capacity of fish stocks. Consequently, places such as the coast of Terranova in Newfoundland, which has been the site of cod fishing for five centuries, are presently deprived of precious fish. By 1992, the sea was empty, yet not even the Canadian government's prohibition of fishing has managed to change this situation, which to this day has remained the same. In many European seas, things are not better. Excessive fishing, denounced in the policy debates of the EU Green Book in March 2001, has led to the collapse of forty of the sixty main stocks of fish in the Northeast Atlantic.[7]

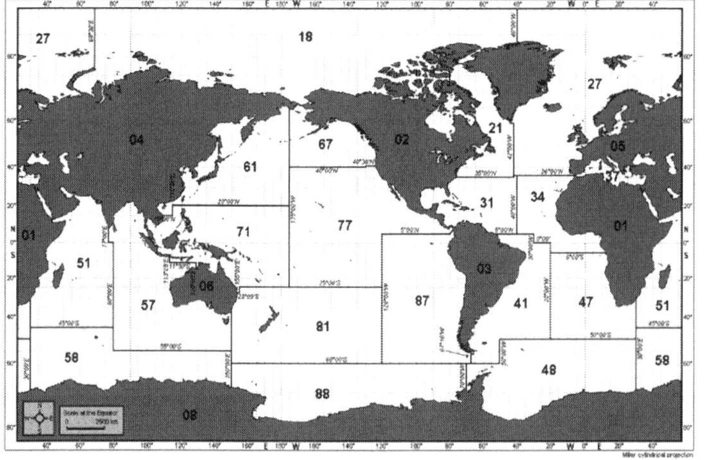

Main fishing areas

01- Africa
02- North America
03 - South America
04 - Asia
05 - Europe
06 - Oceania
08 - Antarctica
18 - Arctic Sea
21 - Atlantic Ocean (North-west)
27 - Atlantic Ocean (North-east)
22 - 32 Subdivision of the Baltic Sea
31 - Atlantic Ocean (Center-west)
34 - Atlantic Ocean (Center-east)
37 - Mediterranean Sea and Black Sea
41 - Atlantic Ocean (south-west)
47 – Atlantic Ocean (south-east)
48 – Atlantic Ocean (Antarctic zone)
51 – Indian Ocean (west)
57 – Indian Ocean (east)
58 - Indian Ocean (Antarctic zone)
61- Pacific ocean (north-west)
67 – Pacific ocean (north-east)
71 – Pacific Ocean (center-west)
77 – Pacific ocean (center-east)
81- Pacific Ocean (south-west
87- Pacific Ocean (south-east)
88 – Pacific Ocean (Antarctic zone)

Source: www.fao.org/fi/maps/world_2003.gif.

At the turn of the millennium, the total production of sea fishing reached 94.8 million tons, the highest level ever registered, but estimates, made on the basis of forecasts by major countries, demonstrate a decisive reduction, amounting to about 92 million tons. If we exclude China (the main producer), global production based on the catch of 2000 has suffered a further loss, 78.million tons compared to 83 million in 1989.[8] From the following figure we can get an idea of the progress of global production.

World Production of Fish Catch in 2000
The areas represented are those in which production has exceeded two million tons.

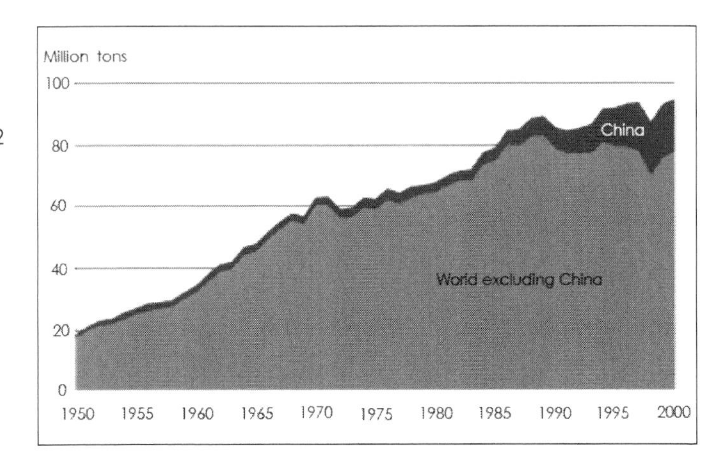

Production of Fish Catch in the Main Fishing Areas in 2000
The areas represented are those in which production has exceeded two million tons.

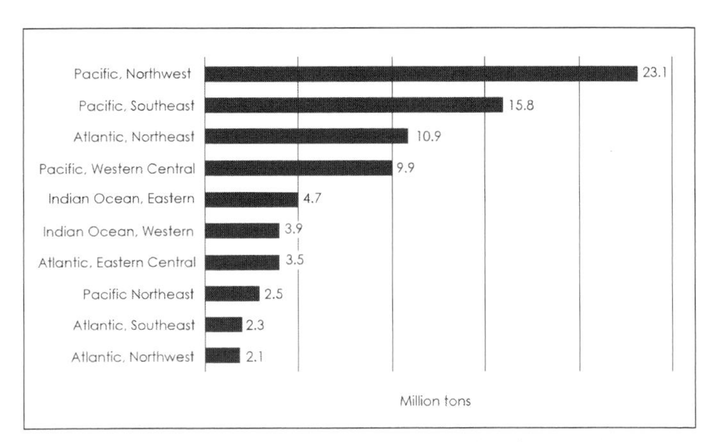

Source: UN Food and Agriculture Organization, "*Situation mondiale des pêches de l'aquaculture*," in *Rapporto SOFIA 2002: La situation mondiale des pêches et de l'aquaculture*, 8–10.

World Fishers and Fish Farmers by Continent

	1970	1980	1990	1991	1992	1993	1994	1995	1996	1997	1998	1999	2000
Total	*in thousands*												
Africa	1360	1553	1917	2092	1757	2032	2070	2238	2359	2357	2453	2591	2585
North America Central America	408	547	767	755	757	777	777	770	776	782	786	788	751
South America	492	543	769	738	763	874	810	814	802	805	798	782	784
Asia	9.301	13.690	23.656	24.707	25.423	26.342	27.317	28.552	28.964	29.136	29.458	29.160	29.509
Europe	682	642	654	928	914	901	881	864	870	837	835	858	821
Oceania	42	62	74	77	79	80	74	76	77	78	82	82	86
World	12.285	17.036	27.837	29.297	29.691	31.005	31.928	33.314	33.847	33.995	34.411	34.163	34.536

Source : Fao, *Situation mondial des pêches de l'aquaculture*, in Rapporto Sofia 2002: *La situation mondial des pêches de l'aquaculture*, p.16.

With this continuous growth, the number of operators has greatly increased. Out of 35 million persons active in the field, the number of fishermen increased by 2.2 percent per year starting in 1990.

In this context, the regional organizations of the continental fishing industry that are responsible to the international community have not succeeded in effectively protecting the stocks most profitable for the market, having more of a consulting role than managing power, which has largely been delegated to national authorities.[9] Let us remember, in this regard, the International Whaling Commission (IWC), founded in 1946, that has not succeeded in controlling whale hunting. In just the few years that followed a moratorium placed on this practice, which came into effect in 1998, Japan has hunted a total of 5,779 whales on the pretext that catching whales has scientific purposes.[10] But the great cetaceans are not used to bolster scientific research, as is officially claimed; instead, they reach international markets, where whale meat is destined for consumption. This type of trade, moreover, has grown. Every year more than 1,300 animals are killed, in violation of the international ban on the trade in whale meat established by the Convention on International Trade in Endangered Species of Wild Fauna and Flora (CITES). Far from complying, Japan and Norway, another country involved in the hunting of cetaceans, would like to see this international trade reopened. In 2002 the Norwegians even decided to increase their quota of whales to be hunted, from 552 specimens to 672. According to the Norwegian whale industry, such an increase would be justified by the size of the total population of rorquals, which amounts to 118,000 animals in the catching areas.[11] In the fifty-fourth meeting of the IWC (May 24, 2002), Japan, not having succeeded in overturning the moratorium, blocked the creation of protected marine areas for whales in the oceans of the South Pacific and in the South Atlantic.[12]

As for the situation of the fishing industry in the European Union, this too is not without problems. The communitarian management of one of the main fishing markets in the world is not a simple thing. Italy, which is in second place among the beneficiaries of European subsidies to the fishing industry, contributes to overfishing with an annual increase in its fleet of

7 percent. The main ports where fishing activities are concentrated are: Naples, Venice, Bari, Mazara del Vallo, Chioggia, and Catania. The total number of ports and landing places on national territory is eight hundred. Despite the fact that 80 percent of the fleet is composed of small boats, generally quite old, modern fishing boats use very sophisticated technologies—satellite navigation systems (global positioning systems, GPS), mapping with sonar—originally developed for military use. In order to better trace schools of fish, echo sounding with colored chart plotters is used, replacing traditional depth sounders that only provide information about the depth of seabeds. The enhanced navigation and security technology increases the possibility of fishing, by increasing the number of days in which it can be carried out over the course of the year.[13]

In conclusion, here is a table that clarifies the Italian situation (see also Appendix One):

Italy	
Number of boats	16,522 (2002)
Total gross tonnage	about 200,000 gross ton
Overall capacity of the Italian fleet	1,300,000 Kw
Number of fishermen	53,000
Employment in the entire sector	70,000
Annual catches	463,400 tons
Value of the catches	2023 million euro
Average annual consumption of fish per person	22 kilos
Internal Production (aquacolture included)	760,000 tons
Subsidies disbursed by the EU	352.50 million euro in the period between 1994 and 1999

ILLEGAL FISHING

By "illegal fishing" we refer to a whole series of activities that, besides being outlawed, seriously damage the environment and the economies of the fishermen who operate within the law. This kind of fishing—with no rules and licenses and using forbidden equipment—concerns the exclusive economic zones and deep-sea fishing, and constitutes a truly global problem. According to FAO estimates, 30 percent of the total fish caught is caught thanks to illegal, undeclared, and unregulated fishing.[14]

The Indonesian coral barriers, the wonders of which we have already described, find 85 percent of their coral at risk because of this kind of fishing, which is carried out with explosives and goes to feed clandestine markets of collectors.[15] In the islands of the archipelago the bombers go to seek schools of fish swimming in the coral forests; they utilize crude bombs (made with a small bottle filled with a mixture of fertilizer and kerosene that is connected to a cheap subaqueous fuse), and after having stunned or killed their prey, easily pick up the loot. On the bottom there remains a pile of coral debris about the size of a car and around it branches and scrolls broken, shattered. These are the ruins of decades, if not centuries, of slow growth. At a national level, this type of practice causes damage of approximately $500,000 per day, and the fleet of bombers, making up to thirty explosions a day, makes the bottom flat and lifeless.[16]

After this dramatic example, let us turn to what occurs in the Mare Nostrum, where, from the Venice lagoon to the strait of Messina, illegal practices boom.[17] One of the most widespread is the use of forbidden nets with mesh that leads to the catching of small fish. Many bottom trawlers, in fact, tend to use nets with meshes smaller than those authorized, or drift nets, making it difficult for small fish to escape. Another illegal means of fishing is trawling within three miles of the coast, on low seabeds or rocky bottoms, and on the sea grass, or *Posidonia*. According to a study by the World Wildlife Federation (WWF), 83 percent of the red tuna and swordfish caught in the Mediterranean is undersized. In some ports of southern Italy they sell swordsfish as small as mackerels, while in order to be traded legally a swordfish should measure at least 1.40 meters in length. The WWF denounces the consumption of baby fish, which is sadly a widespread occurrence in many regions. In Italy, for instance, "*bianchetto*" indicates the small fry of various species of fish, which is transformed into a mush used for fried food and stuffing. In a few grams there are dozens of lives, and a kilo of *bianchetto* means the death of thousands of fish that would have grown and achieved a larger size. In some localities, the "*ceche*" are eaten (small eels) these too fished with nets with mesh as small as that of a mosquito net. The problem does not concern only fish but also crustaceans and mollusks. On the banks you can find, for instance, small dead rock polyps only few weeks old, a tragedy if we think that an adult specimen should weigh a kilo.[18] Tons of small swordfish and tuna are caught and routinely sold at rock-bottom prices in gross retail markets, fish shops, and restaurants. In Italy, certain novelty fish are much in demand because they are part of Neapolitan cuisine, and they are grabbed by any means, even with the famous final "panno" [cloth], and with tile that presses and destroys not only very small fish but recently formed larvae.[19]

Another unresolved problem is also the fishing of date mussels, which has been forbidden since 1988 but continues to destroy the limestone rocks of Italy.[20] The date mussel, a mollusk that has the peculiarity of living inside these rocks and constructing deep galleries, even twenty centimeters down, may spend twenty years building five centimeters; it is considered the most exquisite and precious of the bivalves. To collect it, date mussel fishermen use axes, scalpels, and jackhammers, constantly changing areas in search of new specimens. It is precisely technological improvement and the spread of subaqueous immersions that has increased the number of people who engage in this profitable activity, whose impact is truly destructive. These benthonic rocks, in fact, are among those with the highest level of biodiversity in the Mediterranean, and they constitute a true source of nutrition for several species of fish. The restoration of their equilibrium, making the settlement and growth of new populations possible, requires many years. However, this is not only an ecological catastrophe; the damage is also economic. This indiscriminate fishing, practiced especially on the Sorrento peninsula, particularly on the seabeds of the protected areas of Punta Campanella, on the southeastern coasts of Sicily, on the coasts of Puglia, and in the Cinque Terre, seriously compromises the entire economy of small fishing. The detachment of entire walls of rock desertifies the area and chases away other species of fish. Every year, just on the Amalfi coast, about eighty thousand square meters of rock is eliminated, at a cost equal to about €5.2 million.[21] It is enough to think that during the Christmas holidays, at Easter time, and in the month of August the price of date mussels oscillates between €30 and €75 per kilo.

In the province of Naples there is another worrisome phenomenon: the abusive, illegal sale of mussels in disregard of hygienic and sanitary regulations. Not surprisingly, hepatitis A infection has reached a record level in this area. In the period from December 2001 to May 2002 about forty-five tons of the precious fish product, cultivated and sold illegally and without any hygienic control, was confiscated and destroyed. In the Venice lagoon, as pointed out by Luca Ramacci, then the solicitor general, in the *Dossier Mare Monstrum 2002* of Legambiente (the Italian League for the Environment), the illegal fishing of mollusks is practiced in areas that are highly polluted. While passing over the bridge that connects Mestre to Venice one often sees small boats working just in the vicinity of the industrial dumps of Marghera. The clams fished there, using the most disparate means, are not subjected to any sanitary control and are put on the market through channels parallel to the ordinary ones, utilizing falsified sanitary documents.[22] This activity, besides devastating the seabed, constitutes a true threat to the health of unaware consumers. To get an idea of the habitat in which these clams grow we can consult a table showing the average values of the different pollutants present in the mollusks. This is found on page 120 of the dossier.[23]

To this depressing picture we must add pseudo-sport fishing that is done with professional equipment, without licenses or respect for sanitary norms and fiscal obligations. The "Sunday" fishermen operate, in fact, in the majority of cases, with total disregard for existing laws and good sense. They move several miles off the coast and lower hundreds of fishing lines to great depths. The latter affect the last reserve of fish: the fish fauna living at one hundred to two hundred and even eight hundred meters deep, composed of big groupers and snappers and other species of high value that are sold on the black market. The operations of inflatable boats, pilot boats, motorized fishing boats, and motorboats, with fishing lines armed with small spoons, feathers, and true and rubber fish to fish for swordfish, tuna, small tuna, *albacore*, amberjacks, needlefish, dentexes, and mackerels, have reached considerable proportions.[24] Thus, the General Association Italian Fishing Cooperatives (Associazione Generale Cooperative Italiane per la Pesca, or AGCI) and the WWF consider the regulation of this sector an urgent matter. It would be necessary, then, to put some order in it, with the introduction of an ad hoc regulation, to prevent this situation from deteriorating. It is not a question of banning a hobby but surveying this activity to make it transparent and easily controllable. For those who fish for sport, in fact, there are no reliable data because, unlike their commercial fishing colleagues, no census is taken of them. There could be a million and a half practitioners, but if we also count those who like to fish from the shore and underwater, it could reach up to two or three million.[25]

Unfortunately, calculated in relation to illegal fishing as a whole, the percentage of boats caught red-handed is ridiculously low, compared with those that constantly practice forms of fishing that are banned and destructive for marine species and the income of the fishermen who respect the law.

THE ENVIRONMENTAL IMPACT OF INDUSTRIAL AQUACULTURE

In recent years the situation of aquaculture worldwide has undergone notable changes to cope with the constant increase in fish consumption.[26] By the turn of the millennium, in fact, production had increased by 11 percent yearly, so much so that the blue farms have become the most rapidly growing food-producing economic sector.[27] To better understand this matter, let us analyze a graph constructed by elaborating some data provided by the FAO, which compares the quantity of fish fished in the wild with that of fish farmed between 1972 and 1998.

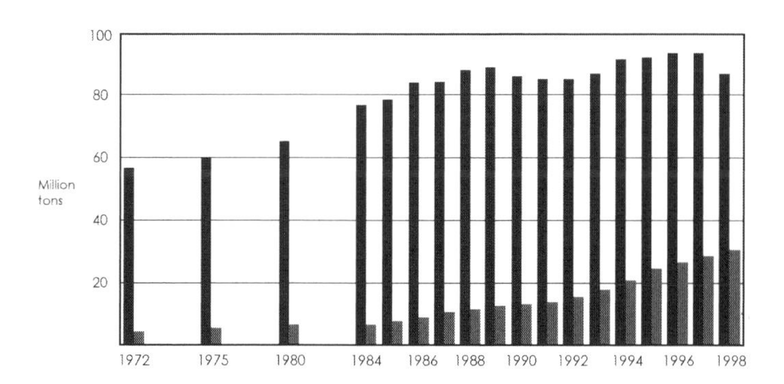

The numbers placed on the ordinate axis represent million of tons of fish; those in pale gray represent those caught with fishing techniques and those in dark gray those raised through aquaculture. On the axis of the abscissas we find the temporal dimension. From the evolution of the histograms we can easily observe the constant growth of activities that are tied to aquaculture—a growth that, though not reported in the grap, has continued also in recent years.

Source: www.acquaguide.com

This growth can also be explained by the great advantages that aquaculture offers with regard to the production of cheap animal proteins. Compared with cattle farms, fish farms need a much smaller quantity of cereals: cattle need seven kilos of cereals to gain a kilo; fish can do it with less than two kilos of fodder. There is, moreover, a remarkable economy of water, because it takes one thousand tons of water to produce a ton of cereal. From the following graph elaborated by the Worldwatch Institute we can see the production of fish and meat between 1950 and 2000.

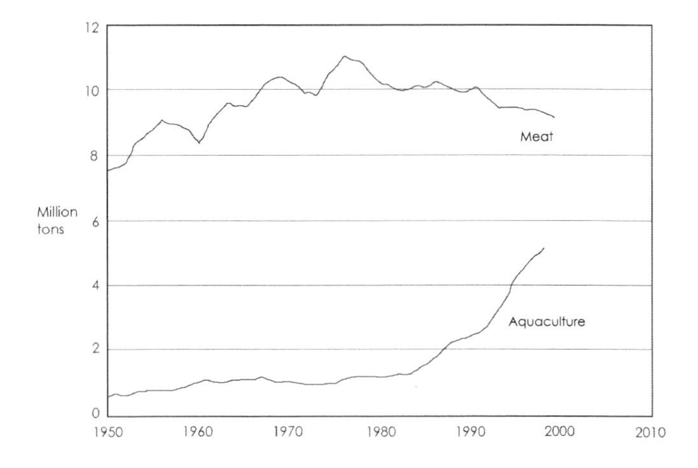

Source: www.wwf.it/ambiente/earthpolicy/acquacoltura.asp.

About 85 percent of this production is concentrated in developing countries. Since the 1970s, in fact, the World Bank has supported this type of activity by giving subsidies to governments in Asia and Latin America for the construction of tubs for the cultivation of shrimp.[28] As for the industrial countries, Japan, the United States, and Norway lead production. In Europe the aquaculture industry has registered constant growth, performing an increasingly significant function, especially in Austria, Italy, Greece, and France.[29] In Italy, fish production concerns above all mollusks, which are 73 percent of the output; there are also excellent prospects for the cultivation of sea fish.[30] However, according to recent estimates the majority of the products that arrive at tables in Italy and across Europe are imported. In the first four months of 2001, more than 400,000 tons of fish were imported to Italy, coming mostly—more than 56 percent—from the countries of the EU (Spain, Denmark, Holland, France, Greece, the United Kingdom, and Germany); the remaining 44 percent (more than 177,000 tons) came from other countries, among them Argentina, Morocco, Thailand, and Cambodia.[31]

Aquaculture, however, produces a whole set of negative interactions with the environment, and possible alterations. Although one fish out of four that we consume comes from blue farms, the reserves of fish coming from the sea are diminishing dreadfully.[32] The aquaculture sector uses more resources than it generates. Just looking at the resource "fish" we find that, according to various scholars (as we will see later), in order to produce fodder we destroy more fish than we produce. The rapid development of intensive farming techniques that make it possible to have more and more intensive cycles of production—the introduction of new species; the use of chemicals and drugs, in particular antibiotics to reduce the risk of pathologies in the farms—raise new problems from the viewpoint of the sustainability of these activities.

Both inshore and offshore installations, as well as farms relying on tubs on land fed with seawater or salty water, raise grave concerns for marine fauna. These in fact take water from the sea and return it later, or take it from underground aquifers and then return it to the sea, to coastal lagoons, or to drainage canals contaminated by feeding materials, chemical substances, and massive doses of fish excrement. Freshwater installations too have an impact on the surrounding environment in ways not always compatible with its regeneration.

These farms, then, undermine the ecological habitat and its natural resources, polluting above all the ecosystems of the coastal areas and lagoons, which are among the most important and sensitive in the world. To cut the initial costs, aquaculture farmers exploit existing natural resources, constructing their farms near areas that are humid, rich in water, and have a terrain that can be used easily, without investing in technologies that would benefit the environment and would be advantageous in the long term. In Asia and South America, for instance, the intensive farming of shrimp has been demonstrably unsustainable, because the construction of the tubs for these crustaceans is the primary cause of the destruction of the mangrove forests. To produce this product, which we may call "luxury food" destined for rich countries, the areas in which the precious plants grow have been irreversibly reduced. Of the 203,765 hectares of mangroves lost in Thailand between 1961 and 1993, 32 percent has been sacrificed to the farming of shrimp.[33] We will analyze these issues in more depth later to explain the disastrous impact that these installations have on developing countries.

Many species fished in the sea, moreover, are often used to feed trout, bass, and farmed salmon. It typically takes two pounds of fish caught in the sea to feed a pound of farmed fish.[34] Another species that, together with shrimp, is increasingly becoming a pretext for environmental destruction is salmon, with a global production of seven hundred thousand tons a year. This type of fish, originally widespread in the northern part of the Atlantic Ocean, is farmed and consumed in industrialized countries, Norway above all. Here, however, it is fed with other sea fish, such as anchovies and herring, or with leftovers from the production of fish, a practice that puts more pressure on oceanic fishing.[35] For every ton of salmon produced, up to five tons of other fish are necessary. Another problem connected with salmon farms is the danger that, due to damaged tubs or attacks by predators such as seals, species farmed for quick growth may escape and possibly cross with wild salmon, reducing the latter's capacity for survival.[36] These "cousins," born in captivity, on leaving the farms cross with wild salmons and generate unfortunate genetic modifications that make the preservation of this specific population difficult. Wild salmon is genetically different depending on the environment and the river in which they find themselves and to which they tend to return. This property makes repopulation very difficult and requires higher attention from the farmers. In this sense, a positive example comes from the

Scottish group Scottish Quality Salmon, which is the leader in the industry that produces salmon and has formed a working group for the purpose of ensuring a healthy balance between farmed and wild salmon.[37] It is this type of farm that degrades the coastal waters of Maine, for example, and according to estimates gathered between the summer and the winter of 1999 by the University of Maine, the levels of oxygen dissolved in the waters are already low in many areas of the state's Blue Hill Bay.[38]

In this sector exist, moreover, various biotechnologies aimed at creating superfish.[39] These genetic modification techniques raise the interest of the fishing industry, because they can make fish grow more rapidly and resist low temperatures. Specimens of salmon that can survive frost have been engineered from natural species that cannot survive freezing temperatures by genetic grafting from arctic sole, which prevents the formation of crystals in the blood.[40] In the United States and Canada, moreover, the antifrost protein has been used to engineer an Atlantic salmon with a gene for hormonal growth that makes it possible to trade the fish in twelve to eighteen months rather than in the canonical three years.[41] In Europe, as well, the only transgenic animal recently patented is a kind of salmon that can reach dimensions several times superior to those of the wild species.[42] Genetic manipulations, however, offer a troublesome scenario, as the crossing of laboratory species with wild ones has led to a loss of diversity and the risk of negative consequences, still little known, in the short, medium, and long term.[43] According to bioengineers, genetic modification would have indubitable benefits. Among the last claimed is the possibility of discarding the use of antibiotics. To create fish resistant to diseases that in intensive farms are easily transmitted, leading to the explosion of violent epidemics, would seem to be the only alternative to the use of drugs.[44]

Indeed, antibiotics are used abundantly in aquaculture to treat the infections caused by various pathogenic bacteria. But concern for the consequences of this practice is growing, as the drugs administered, mixed with food, are put in the water, putting pressure on the exposed ecosystem. There is also the problem of resistance to antibiotics. The *Aeromonas salmonicida* is an example of a pathogen of fish that in many regions is frequently resistant to drugs commonly used in aquaculture, such as sulfonamides, tetracycline, amoxicillin, trimethoprim-sulfadimetossina, and quinolones.

The FAO has raised this question in its report "The State of World Fisheries and Aquaculture," stressing the effects of the industry and its dangers for public health.[45] Indeed, antibiotics, when they are not used as medicines, can have secondary effects that can be avoided if one respects the recommended dosage and the duration of the therapy. By contrast, when they are taken by unaware consumers in the form of residues in alimentary products, they cannot be quantified and have direct health consequences. Aplastic anemia, for instance, is associated with the antibiotic chloramphenicol.[46] The unintentional ingestion of antibiotics promotes a resistance to the latter, but also in bacteria that are pathogenic for human beings. These worrisome features have led to the banning of some of these drugs and the fixing of maximum limits for the residue.[47] Controls, however, are not always capable of guaranteeing that the product is safe. Actually, in 2002, the presence of chloramphenicol in shrimp sold on the international market raised many questions. This substance, found in products of aquaculture, caused the slowing down of imports, with negative economic consequences for the whole market.[48]

Moreover, fish grown in cages in the sea, in enclosures or tubs, produce a great quantity of waste. The farmers who manage intensive installations, where the quantity of the fish fauna in the tubs is very high, must worry, above all, about the waste coming from the cycle of production. One of the main problems is posed by the organic substances deriving from excrement, excessive fodder, urine, silt, and the metabolic process. Dissolved, these substance lead to an increase in the concentration of ammonia, nitrates, and phosphorous, changing the organoleptic characteristics of the water. The presence of colloidal substances—of an organic nature or otherwise—forms, moreover, a foam that rises to the surface, producing the phenomenon called floatation. The elimination of these water dumps is a complex problem if adequate treatment plants are not used. The very disposal of dead fish, considered high-risk waste insofar as they are carriers of pathologies, must take place through specifically authorized installations.[49]

As for salmon grown in Norway—another important example of unsustainability—their waste is equal to that produced by four million Norwegians.

Other threats to the health of the Italian consumer, caused above all by imported products, are identified in Legambiente's 2002 *Dossier Mare Monstrum*. In non-European countries, in fact, there are fewer controls, and the following are some consequences:

- Antibiotics are used in the larval stage of the fish, and consequently end up on the plate of the consumer;
- To disinfestations the tubs, toxic substances are used, or even carcinogenic ones, such as furaltadone and furazolidone (potentially toxic substances), malachite green (harmful following acute exposure, it presents serious risks if ingested, inhaled, or brought into contact with skin), and formalin. In this case as well, the food chain brings these substances directly from the flesh of the fish to the plates of those who eat it;
- In mariculture installations, antifouling varnishes are used to protect nets. These varnishes contain tin and other heavy metals and PCBs (polychlorinated biphenyls) that are carcinogenic;
- Low-quality feed is used, in some cases with concentrations of poisons.[50]

In Italy, on the contrary, fish production is self-controlled by the fishermen's cooperatives, through risk analysis and the control of critical points (hazard analysis and critical control points, HACCP); periodically, the installations are inspected by the local public health agency (Azienda Sanitaria Locale, ASL), and by the Nuclei Antisofisticazioni e Sanita dei Carabinieri (NAS), a police body in charge of supervising food products and preventing food adulteration.

One more source of concern for environmentalists is the boom in offshore farms, floating cages that endanger the red tuna stocks of the Mediterranean that are already impoverished by industrial fishing. The tuna fished alive are placed in these cages to fatten, to make them more desirable for the Japanese sushi market, but with methods that provoke strong organic pollution. Japanese buyers can pay tens, sometimes hundreds of thousands, of dollars for one tuna.[51] This practice, still little known, has expanded in recent years, extending to Spain (the most productive installations in the world are along the Murcia coast), Malta, Tunisia, Algeria, France, Italy, and Croatia. The twelve existing installations produce eleven thousand tons of tuna, more than half the quota farmed in the rest of the world at the time of this writing, without any control by the responsible organizations, namely the General Fisheries Commission for the Mediterranean (GFCM) and the International Commission for the Conservation of Atlantic Tunas (ICCAT). The environmental impact of this activity is particularly worrisome, because breeding occurs during the summer months, when the temperature of the water is higher, and often in tourist areas.[52] Recently, the Croatian island of Vis has opposed, with a referendum, a plan to build new cages offshore near its coast. The decisive victory (88 percent of the inhabitants voted against it) has thus prevented an alteration of the surrounding landscape, contributing to the preservation of the tuna stock of the Adriatic Sea.[53]

Against this type of installation there is a mobilization of Mediterranean environmental associations that have proposed a moratorium on their development.

POLLUTION

To irrational fishing practices, and to the impact of aquaculture, we must add the increasing impoverishment of resources due to all the industrial, agricultural, and domestic activities that send their waste into waters flowing toward the sea.[54] The causes of the degradation of the water are manifold both in kind and origin. Polluting elements can be differentiated into biological, chemical, and physical, and on the basis of their origin we can classify them as follows: urban and domestic liquids; industrial waste; wastewater from farms; pollution from agricultural runoff, due to the intensive use of chemical products (fertilizers, herbicides, pesticides); and pollution due to navigation and commercial transportation.

Bacteriological pollution of fecal origin—fecal coliforms and streptococcus—and organic pollution—basically made of carbohydrates, proteins, fats, detergents, phenols, and amines—show the civil contributions to the anthropic pressure on the coasts. In Italy there has been an attempt to limit such flows, thanks to the increasing adjustment of civic and industrial discharges imposed by a permit system set up by the Merli Law (L.319/76). Even today, however, one can find the presence of untreated or illegal domestic discharges on the coastline of very urbanized areas and within ports. These pollutants can lead to eutrophication in coastal zones, with incalculable damage to ecosystems—the destruction of benthonic communities; the death of numerous fish—and to the economy and tourism. Although the problem of eutrophication is rather complex, the principles that cause it are fairly clear. The increase in primary production (phytoplankton) is caused by an increase in the concentration of dissolved nutrients. Phosphorous and nitrogen are among the main elements responsible for the eutrophy of the waters.[55] In the 1980s the phenomenon reached remarkable dimensions in the Adriatic Sea, with repercussions all along the coast.[56]

There is a growing attention also to the impact of so-called persistent bioaccumulative and toxic molecules (PBTs). Among them we find chlorinated hydrocarbons. The most famous is DDT (dichloro-diphenyl-trichloroethane), forbidden since 1972. In the past it was widely used, and the effects of its presence are felt even today. In this category there are also polychlorinated biphenyls (PCBs), persistent industrial compounds that were produced for about thirty-five years before being identified as environmental pollutants. In recent decades there has also been an increase in heavy metals deriving from industrial activity, traffic, and the growth of waste. The metals that can be found accumulated in mussels, for instance, are cadmium, lead, zinc, and nickel.[57] Concerning the accumulation of these substances in the fish of our seas, recent estimates seem reassuring, as the level of toxic substances is rather low, obviously with the exception of the zones close to industrial plants. Taking into account the fact that the threshold of potential risk for dioxin and other equivalent toxic substances is fixed at between 1 and 4 picograms per gram, we now consider data from an analysis of Italian fish. Here are the average values, calculated in picograms per grams (1pg = 10^{12}g), of the equivalent toxic substances (Teq) in various fishing products:

Ligurian Sea		**Low Tyrrehnean**	
Hake	0,50 pg/g p.f.	Hake	0,71 pg/g p.f.
Sardine	0,80 pg/g p.f.	Sardine	0,19 pg/g p.f.
Shrimp	0,20 pg/g p.f.	Shrimp	0,36 pg/g p.f.
Mullet	0,50 pg/g p.f.	Mullet	0,51 pg/g p.f.
Octopus	0,06 pg/g p.f.	Octopus	0,45 pg/g p.f.
High Adriatic		**Ionian Sea**	
Hake	0,45,pg/g p.f.	Hake	0,36 pg/g p.f.
Shrimp	0,16 pg/g p.f.	Sardine	0,20 pg/g p.f.
Sardine	0,99 pg/g p.f.	Shrimp	0,12 pg/g p.f.
Mullet	0,35 pg/g p.f.	Mullet	0,33 pg/g p.f.
Octopus	0,19 pg/g p.f.	Octopus	0,08 pg/g p.f.
Low Adriatic		**Southwestern Sardinia**	
Hake	0,46 pg/g p.f.	Hake	0,41 pg/g p.f.
Sardine	0,30 pg/g p.f.	Sardine	0,79 pg/g p.f.
Shrimp	0,15 pg/g p.f.	Shrimp	0,13 pg/g p.f.
Mullet	0,36 pg/g p.f.	Mullet	0,27 pg/g p.f.
Octopus	0,22 pg/g p.f.	Octopus	0,19 pg/g p.f.

In all these cases, and for all the observed species, even the worst values have remained beneath the potential threshold of danger for humans. However, the presence of polluting substances, although below the parameters imposed by the EU, underlines the importance and necessity of interventions aimed at improving the state of health of the sea and guaranteeing a higher level of alimentary and sanitary safety for consumers.

Source: Legambiente, *Dossier Mare Monstrum* (2001), 86, available at: www.Legambiente.com/documenti/2001/06DossierMareMonstrum/intromaremonstrum.html.

The northern seas, from this viewpoint, are more at risk.[58] The Baltic Sea, in particular, is a very critical zone because of pollution caused by dioxin, a highly toxic organic substance, due to the numerous metal, extractive, and fertilizer industries built next to the sea.[59]

To the pollutants that come from the rivers we must add those that derive from navigation and sea accidents. The latter, when they involve tankers, cause catastrophes whose consequences are incalculable. The disaster that put Galicia on its knees in the fall of 2002 is one of many accidents that, in the course of history, have devastated entire ecosystems. According to the Spanish WWF, the damage to local economies dependent on fishing and mariculture amounts in this case to €100 million.[60] Ships such as the *Prestige*, an oil tanker than sunk off the coast of Galicia in 2002, should simply not be permitted to navigate the seas. Instead, ignoring existing international conventions—MARIPOL on marine pollution and SOLAS on navigation safety on the sea—these wrecks of the sea are true ecological bombs. The Mediterranean, being a closed sea, is particularly at risk, above all because a fourth of the world's traffic is concentrated in its basin.[61]

On top of everything else hovers global warming and the consequent climate changes, which constitute further concerns for already compromised fish resources.

NOTES

1. H. Jonas, *Il principio responsabilitá. Un'etica per la civiltá tecnologica* (Torino: Einaudi, 1990).
2. This is how Jules Verne defined it in its *Twenty Thousand Leagues Under the Sea*, in 1870: "The sea is the vast reservoir of nature. The globe began with sea, so to speak, and who knows if it will not end with it?" (J. Verne, *Twenty Thousand Leagues Under the Sea*, in *Omnibus Jules Verne*, 58, accessed at: www.netlibrary/com/reader).

3. The first fishing boat that could freeze its catch was built in 1953 for the Salvesen's, a company engaged in whale hunting in the waters of the Antarctic (A. Caffarena, *Governare le onde*, 33).
4. M. Carbone, *Le milieu marin.*
5. UNCLOS, United Nations Convention on the Law of the Sea, articles 55–75. The last was introduced after the Third United Nations Conference on the Law of the Sea (UNCLOS III), which ended with the drafting of the Convention on the Law of the Sea. The convention, which was ready for signing on December 10, 1982, came into effect only in 1994. It represents one of the most significant provisions in the governance of marine resources, as it places 40 percent of the seas under the jurisdiction of coastal nation-states. [per Web site: http://legal.un.org/diplomaticconferences/lawofthesea-1982/lawofthesea-1982.html]

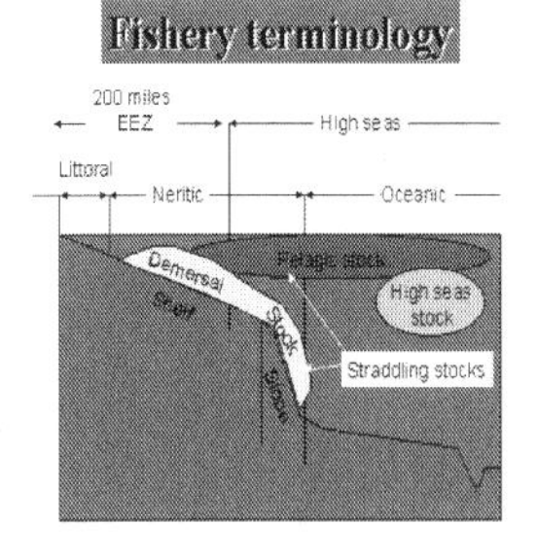

Source: www.oceanatlas.com/world_fisheries_and_aquaculture/htmlgovern/capture/highseas/img012.gif.

6. The politics of enclosure, like the institution of EEZs (exclusive economic zones), have not limited the irrational exploitation of the sea. (B. S. Frey, *Economia politica internazionale* [Milano: FrancoAngeli, 1987]; J. Rees, *Naural Resources, Allocation, Economics and Policy* [London: Routledge, 1990]; P. M. Wijkman, "Managing the Global Commons" n. 3, *International Organization* [1982])
7. *Stop alla pesca eccessiva o la pesca finirá*, March 8, 2002, available at: www.wwf.it/news/832002_8173.asp.
8. UN Food and Agriculture Organization, "*Situation mondiale des pêches de l'aquaculture*," in *Rapporto SOFIA 2002: La situation mondiale des pêches et de l'aquaculture*, available at: www. fao.org/docrep/005/y7300f/y7300f00.htm.
9. Since 1930 there has been a proliferation of commissions and bodies for the management of fishing areas and the establishment

of quotas for fish and marine mammals. The first of these bodies was the commission for the sea lions of the North Pacific that were hunted for their fur, which was founded in 1911 (N. Meyers ed., *Atlante di Gaia*).

10. *Caccia alle balene riaperta in Giappone*, March 11, 2002, available at: www.wwf.it/news/1132002_1081.asp.
11. *La Norvegia aumenta la sua quota balene cacciabili*, March–April 2002, available at: www.federcoopesca.it/Corriere/n_53_anno_xvi/notizie_flash.htm.
12. *Stop al massacro delle balene*, March 24, 2002, available at: www.wwf.it/news/2452002_4654.asp.
13. www.wwf.it/pesca.italia.asp.
14. *Illegal, Unreported and Unregulated (IUU) Fishing*, available at: www.oceanatlas.com/world_fisheries_and_aquaculture/html/issues/govern/iuu/default /html.
15. *La scomparsa dei coralli*, February 20, 2003, in www.wwf.it/news/2022003_3602.asp.
16. J. C. Ryan, *Battaglia nei mari del corallo*, August 2, 2001, available at: www. enel.it/it/enel/magazine/boiler/boiler52/html/articoli/focusWW-coralli.asp. In addition to illegal fishing, the treasure of biodiversity, in particular that of Central America, is threatened by climate change. “Excessive heat causes the discoloring of corals, a consequence of the disappearance of the minuscule unicellular plants that usually colonize the external parts of the corals, giving them the typical coloring. Without these organisms corals cannot continue to live any longer. According to a report by WWF, between 1979 and 1990, at least 60 cases have been discovered of loss of pigmentation, compared to only three cases reported in the previous one hundred years. Presently 10% of the coral barriers have already been destroyed.” (*Barriere preziose per la biodiversitá*, February 22, 2003, available at: www.wwf.it/news/2222003_1220.asp.)
17. The following is a classification of outlaw fishing in Italy in 2001. The arrows indicate the course of the violations compared with the previous years.

	Region	Confirmed Infractions	Peopl Denounced and Arrested	Confiscations Made
1	Sicily	1707	218	701
2	Puglia	1127	248	783
3	Campania	643	189	309
4	Marche	623	35	514
5	Calabria	587	111	213
6	Tuscany	525	17	396
7	Veneto	411	118	458
8	Lazio	390	101	283
9	Sardinia	317	8	1.656
10	Emilia Romagna	274	91	107

11	Liguria	235	14	78
12	Abruzzo	166	4	123
13	Molise	123	0	99
14	Friuli V. G.	77	9	47
15	Basilicata	2	1	2
	Total	7207	1164	5769

Source: Legambiente, on the basis of estimates provided by customs officers, state and regional forestry authorities, and the ports' harbor offices. Legambiente, *Dossier Mare Monstrum* (2002), 106, available at: www.legambiente.com/documenti/2002/0627DossierMareMonstrum2002/DossierMareMonstrum2002.html.

18. *Per l'estate il WWF lancia il "pesciometro,"* August 8, 2000, in www.wwf.it/news/1832002_1427.asp.

19. Legambiente, *Dossier Mare Monstrum* (2002), 111, availabe at: www.legambiente.com/documenti/2002/0627DossierMareMonstrum2002/dossierMareMonstrum2002.html

20. With Decree-Law No.401 in August 1988, the fishing, consumption, and trade of date mussels were forbidden. Since then, the decree of October 16, 1998 has extended the ban, and the minister for agricultural and forest policies has stressed that imports are equally punishable. But despite the bans, between 80 and 180 tons of date mussels continue to be collected, the equivalent of 6–15 million individuals and 4–10 hectares of seabeds desertified. The figures of the disaster: 15–25 kilos are appropriated daily by a professional date-mussels picker; 500 kilos, the daily appropriation of date mussels along the peninsula of Sorrento; 30,000 square meters of seabeds desertified by date-mussel gatherers every year in the Salento; 70,000 square meters of seabeds desertified every year along the peninsula of Sorrento; €2 million, the yearly business of the date-mussels traders in the peninsula of Sorrento; 1,000 square centimeters, the dimension of the area destroyed for the sake of a plate of a typical kind of pasta, linguine, that is served with date mussels. (Legambiente, *Dossier Mare Monstrum*, 107).

21. *SOS dattero di mare*, May 17, 2002, available at: www. wwf.it/news/1752002_1491.asp.

22. "This activity takes place onboard boats that use different techniques. Once they used the system of the c.d. turbocharger that sucked the mollusks from the seabed. Presently they use a similar method called the vibrating rake, or the technique of the 'rasca' or the 'giostra,' carried out with small boats. These small boats, with powerful outboard engines and radars, are modified through the affixing of a wishbone with lateral supports, on which two other motors are mounted in such a way that the propellers can turn touching the bottom and move the sediment of the lagoon. An iron cage dragged by the boat serves to pick up the clams." Legambiente, *Dossier Mare Monstrum*, 121.

23. Ibid.

24. *Gli impatti della pesca sportiva*, July 16, 2002, available at: www.wwf.it/news/1672002_5979.asp.

25. *Pescatori e Wwf uniti per salvare il mare*, November 13, 2002, available at: www.wwf.it/news/13112002.7987.asp.

26. Seventy-one percent of the global fish population is destined for human consumption, while the remaining 29 percent is used for the production of farines, oils, etc. The following figures represent the world consumption of fish products: nonalimentary consumption: 29 percent; consumption of fresh fish, 26 percent; of salted and smoked fish, 10 percent; of preserved, 12 percent; frozen, 23 percent. (N. Pasella, *L'andamento dei consumi e l'importanza del prodotto*, February 15, 2002, available at: www. aquaguide.com/Partner/Pasella/consumi/consumi.shtml)

27. L. R.Brown, *Alleveremo piú pesce che bestiame?*, s.d., in www.wwf.it/ambiente/earthpolicy/aquacultura.asp.

28. V. Shiva, *Stolen Harvest.*

29. N. Pasella, *Produzione ittica della Unione europea: incostanza della pesca e continua crescita del comparto agricoltura.* November 29, 2001, available at: www. aquaguide.com/partner/Pasella/produzione/prod_europea.shtml.

30. N. Pasella, *Con il 73% dell' output i molluschi occupano un ruolo determinante nell-aquacoltura italiana: ottime prospettive per i pesci marini*, January 7, 2002, available at: www.acquaguide.com/Partner/Pasella/molluschi/molluschi.shtml.

31. Origin of the main kinds of aquaculture imports: bass and giltheads from Greece, Turkey, Malta, and Tunisia; salmon from Norway, Scotland, and Chile; mussels from Spain and Greece; oysters from France; mollusks from Greece and Albania (Legambiente, *Dossier Mare Monstrum*, 129, available at: www.Legambiente.com/documenti/2002/0627DossierMareMonstrum2002/DossierMareMonstrum2002.html.

32. F. Ungaro, *Il rischio dell'aquacoltura.,* October 23, 2002, available at: www.enel.it/it/magazine/boiler/boiler30/html/articoli/AaasUngaro=Aquacoltura.asp.

33. V. Shiva, *Stolen Harvest.*

34. F. Ungaro, *Il rischio dell'acquacoltura*, October 23, 2002, available at: www.enel.it/boiler/boiler30/html/articoli/AaasUngaro-Aquacoltura.asp.

35. The diet of salmon is comprised of 43 percent proteins and 35 percent oils; soy and corn are also items of primary importance for their diet.

36. In 2002, at Kirkwall, the chief town of the Orkney Islands, an archipelago north of Scotland, a hundred thousand salmon poured into the port, because due to the tide, the nets of one of the farms broke. This endangered the indigenous species and caused an economic loss for the owner of the aquaculture

farm calculated as €1.6 million. www.corriere.it/Primo_Piano/Esteri/2002/04_Aprile/02/SALMONI.shtml.

37. S. Ficorilli, *Salmoni in estinzione*, June 14, 2001, available at: www.enel.it/magazine/boiler/arretrati/boiler45/html/articoli/Ficorilli-Salmoni.asp.

38. *L'Acquacoltura rappresenta un potenziale pericolo per l'Ambiente marino*, February 18, 2002, in: www.acquaguide.com/cgi-bin/news/archives.cgi?category=i&view=2.02.

39. The first transgenic fish containing the human growth gene is a rovella, a fish created in China in 1986.

40. S. Morandi, *La fabbrica degli animali*, May 10, 2001, available at: www.enel.it/magazine/boiler/arretrati/boiler40/html/articoli/focusMorandi-Animali.asp.

41. V. Shiva, *Stolen Harvest*.

42. The requests that have arrived at the Europe Patent Office are as follows: 2,000 on human genes (300 were granted before 1998); 600 on animal species (a dozen have been approved); 1,500 on vegetable species (more than 100 approved). In the world, 127,000 human genes have been patented; 29 percent of these belong to one company. (M. Panzacchi, ed., *Organismi transgenici: caratteristiche, rischi, problemi, utilitá*, 2002, available at: www.legambiente.com/documenti/2002/2002ogm.pdf). The first of this type of patent was granted in 1988 by the U.S. Patent Office to DuPont, for a mouse onto which human and chicken genes were grafted; they were infected so they could produce cancer. (V. Shiva, *Patents, Myths and Reality* [India: Penguin, 2001]).

43. In Europe, Directive 2001/18/CE, concerning the introduction of genetically modified organisms and their marketing, is inspired by a cautionary principle; it makes it compulsory, prior to the introduction of any new item in the market, to assess the environmental risk in the short, medium, and long term. (M. Panzacchi, ed., *Organismi transgenici*).

44. S. Ficorilli, *Salmoni in estinzione,* June 14, 2001, available at: www.enel.it/magaziine/boiler/arretrati/boiler45/html/articoli/Ficorilli-Salmoni.asp.

45. As far as resistance to antibiotics, in Ecuador the transference of the latter to pathogenic bacteria occurred during the cholera epidemic in Latin America beginning in 1991. Although the original strain of *Vibrio cholera 01* was sensitive to twelve of the antibiotic agents tested, along the coast of Ecuador the same strain had become resistant to many drugs. This epidemic began precisely among people working in the shrimp farms, where resistance to many drugs was present in noncholeric *Vibrio*s pathogenic for the shrimp themselves. Resistance may have been transferred to *Vibrio cholera 01* by other *Vibrio*s and may have conferred a selective advantage because of the local politics of

chemoprophylaxis. (F. Angulo, *Antibiotici in aquicoltura: potenziale impatto sulla salute pubblica*, Alliance for the Prudent Use of Antibiotics (APUA) newsletter 2000, available at: www.farmacovi gilanza.org/apua/pubblicazioni/20011031.02.asp).

46. Aplasia of hemopoietic tissues is a defect in the regeneration of the hemopoietic matrix that strikes one or more cellular series, or even a whole tissue; and can consist in a numeric diminution of the progenitor cells, down to their complete disappearance, or in a sharp slowing down of their processes of differentiation and multiplication. It manifests itself with peripheral cytopenia.

47. A Maximum Residue Limit (LMR), according to Annex 1 of the European Economic Community (EEC) Regulation n. 2377/90, is "the maximum concentration of residue deriving from the use of a veterinary medical product that the Community can allow to be legally accepted or recognized as acceptable in food." There are substances for which no LMR is assigned because they are not toxic or have a mild pharmaceutical action; these are included in Annex II of the same regulation. In Annex III are included the substances under scrutiny. Annex IV of the same regulation reports all the substances (chloramphenicol, furazolidone, dimetridazole) for which no LMR can be assigned because their residues must be considered dangerous for consumers in any concentration. Here are the veterinary pharmaceutical products authorized for use in aquaculture:

Drug	**Lmr**
tetracycline	100μg/kg
oxytetracycline	100μg/kg
clortetetracicline	100μg/kg
amoxicillin	50μg/kg
flumequine	600μg/kg
florfenicol	1000μg/kg
trimethoprim	50μg/kg
teflubenzuron	500μg/kg
sarafloxacina	30μg/kg

Source: ICRAM, *Linee guida per l'applicazione del regolamento EMAS al settore piscicoltura*, 15/2002/, 87, available at: www.icram.org/documenti/testo linee guida Emas piscicoltura.pdf). For a further analysis of the LMR, compare to *Rapporto* (Sofia 2002).

48. FAO, *Problèmes auxquels sont confrontés pêcheurs et acquaculteurs*, in *Rapporto* (Sofia 2002): *La situation mondiale des pêches e de l'acquacolture*, available at: www.fao.org/docrep/fao/005/y7300f/ y7300f02.pdf. In March 2002 the EU banned the import of shrimp, oysters, and some fish farmed in China, because they contained chloramphenicol (F. Gioannetto, "La rivoluzione azzurra di Sonora," *Il Manifesto*, December 7, 2002; available at: www.ilmanifesto.it/php3/ric_view.php3?page=/terra terra/archivio/2002/Dicembre/3df24ea71cee7.html).

49. In Italy this disposal is regulated by the D.LGS. 508/92.

50. Legambiente, *Dossier Mare Monstrum*, 2002, p.129, in: www.legambiente.com/documenti/2002/0627DossierMare Monstrum

2002/DossierMare Monstrum2002.html.

51. *Boom di allevamenti, a rischio il tonno rosso*, April 30, 2002, available at: www.wwf.it/news/3042002_1212.asp.

52. S. Franchi, *Fattorie blu*, s.d., available at: www.marevivo.it/maric.html.

53. *La Guerra del tonno in Croazia*, February 7, 2003, available at: www.wwf.it/news/722003_9198.asp.

54. Across the planet there are about fifty-eight "dead zones"; they go from the Australian coasts to the Japanese, from the Baltic to the Gulf of Mexico. They are seawaters that are lifeless, where even plankton and bacterial life is missing. They develop when quantities of industrial pesticides and synthetic agrochemical fertilizers are dumped from the rivers into the seas, depriving the water of oxygen. Some fish manage to escape, but the main part of the flora and fauna remain victims of the invaders. The worst zone is in the Baltic Sea. Here more than ninety thousand square kilometers compose a graveyard of red and yellow spots. (M. Forti, "Le zone morte," *Il Manifesto*, June 6, 2000, available at: www.ilmanifesto.it/php3/ric_views.php3?page=/terraterra/archivio/2000/Giugno/3b289437cd140html&word=gamberi).

55.

> The behaviors of phosphorus and nitrogen are fundamentally different. Phosphorous, present in waters in the form of phosphate, at the end of its cycle can be immobilized in sediments through the formation of insoluble complexes, in particular with calcium and oxidized iron. In the case of the absence or lack of oxygen, anoxia, or hypoxia at the level of the water-sediments interface, the phosphorous can be released and return to solution; it can therefore become available again for primary producers (phytoplankton). The cycle of nitrogen is more complex. Nitrogen can enter and exit the system in the form of ammonia gas. Moreover, this nutrient can be removed from the water column, thanks to the process of nitrification and denitrification through which nitrates are reduced by denitrifying bacteria to nitrogen gas that returns to the atmosphere, reducing the quantity of nitrogen available for primary production. Counteracting the eutrophication process is, therefore, necessary to limit the flowing of these nutrients, coming from catchment basins, into the sea to reduce the overall load of both nitrogen and phosphorous that reaches the watery coastal bodies.

Ministero dell'Ambiente, *Tutela dell'ambiente marino* (2000), 269, available at: www.minambiente.it/Sito/settori_azione/sdm/pubblicazioni/qualità_ambienti_marini/docs/conclusioni.pdf).

56. www.minambiente.it/Sito/settori_azione/sdm/pubblicazioni/qualità_ambienti_marini/docs/conclusioni.pdf.

57. Ibid.

58. Legambiente, *Dossier Mare Monstrum* (2001) available at: www.legambiente.com/documenti/2001/06DossierMareMonstrum/intromaremonstrum.html.

59. www.corriere.it/Primo_Piano/Scienze_e_Tecnologie/2002/II_Novembre/12/diossina.shtml.

60. *Petroliere, ancora un disastro ecologico*, November 18, 2002, available at: ww.wwf.it/news/18112002_4044.asp.

61. *Petroliere, Mediterraneo ad alto rischio,* November 18, 2002, available at: ww.wwf.it/news/18112002_6148.asp.

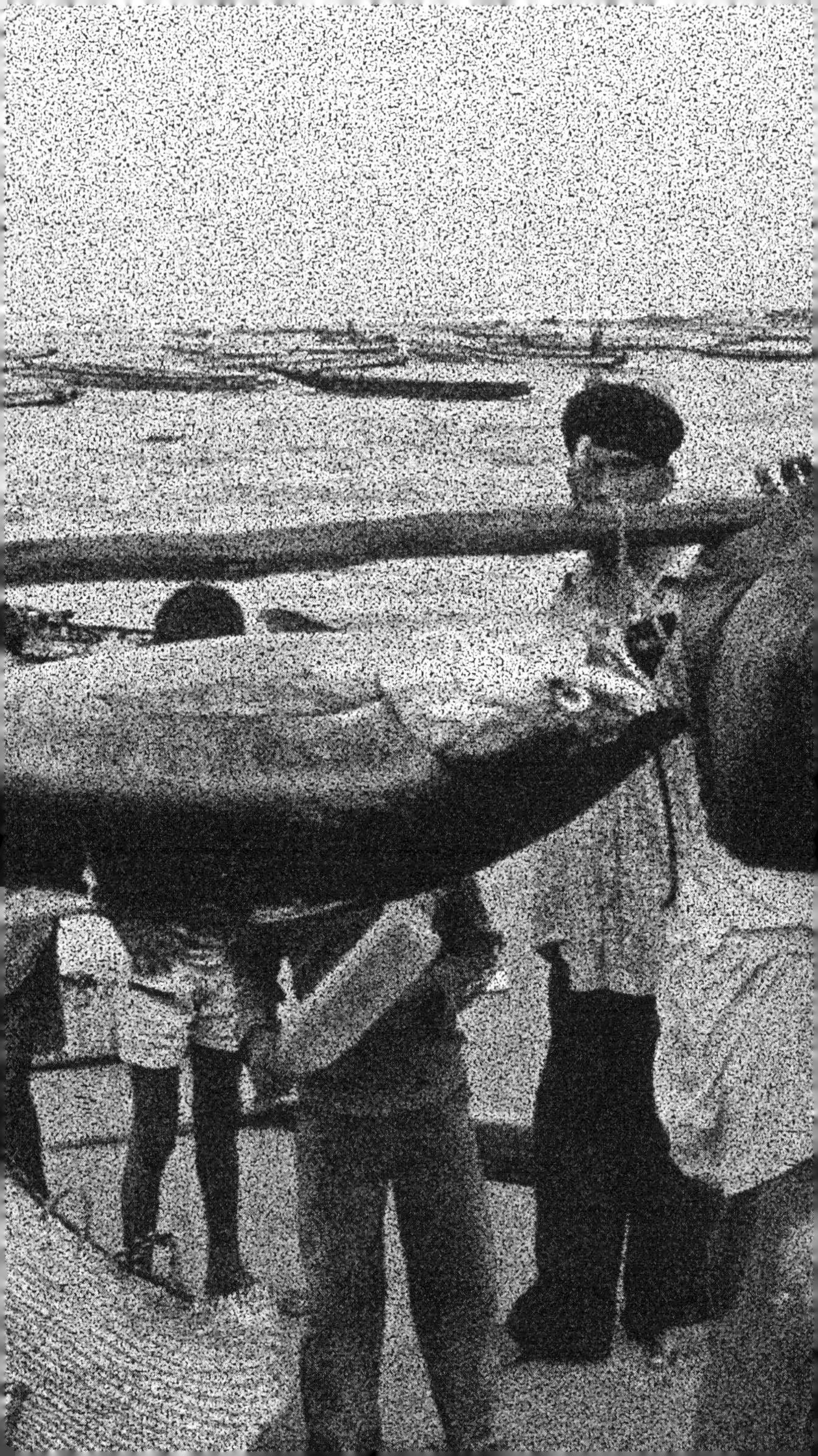

THREE

NEITHER FISH NOR FISHERMEN

Monica Chilese

On the basis of our analysis, let us now consider the state of the resources available and the species that are most at risk worldwide, keeping in mind the hardships experienced by the operators, above all the fishermen who even today continue to practice artisan fishing.

ALWAYS LESS FISH IN THE NETS

A still mechanistic view of nature that conceives of it as a passive, subjugated body has drastically enclosed the vast commons of our seas.[1] The species that have a high commercial value are in fact exploited beyond their capacity to reproduce themselves.

According to the FAO report *SOFIA 2002*, about 47 percent of the main fish stocks or groups are fully exploited and have consequently reached their minimum, or are close to it. Thus, almost half of world marine stocks do not offer any promise of expansion. Eighteen percent are already overexploited and constantly diminishing, without any prospect of expansion, while 10 percent are nearly exhausted. Only 25 percent of fish species are not subject to irrational capture, and the FAO insists that if we do not take adequate measures to reduce excessive fishing, what can be fished will continue to diminish.[2]

Among the groups at risk there are, for instance, many species of sharks that, according to estimates by experts, since the 1970s have registered a 20 percent decline across the world. Fished primarily for their fins and meat, they feed a trade that is constantly expanding, so much so that some of the species, such as the whale shark, the elephant shark, and the great white shark, have also been included in the threatened animals category in the Red List of the International Union for Conservation of Nature (IUCN).[3]

In Europe, the evolution of fishing stocks, from the beginning of the 1970s to more recent years, can be summarized as follows: the majority of round fish stocks (haddock, whiting, cod, hake) have been reduced, and for the most part the volume of the catch is not sustainable; various stocks of flat fish (plaice, monkfish, yellow diamond, sole, mullet) are exploited excessively, but for others the level of exploitation is almost sustainable; pelagic species and those that are the object of industrial exploitation are in better condition, but the volume of the catch must be maintained at the present level or must be reduced to guarantee its sustainability. Various pelagic species show signs of overexploitation, and some may have already reached critical levels; in general, a reduction of exploitation for the majority of the stocks would have favorable economic and ecological consequences.[4]

The scientists of the International National Council for the Exploration of the Sea (ICES) recently asked that fishing be stopped in the seas of the north. The stock of cod, in fact, has been reduced to the bone, and the thousands of fishermen who depend on this resource risk losing their jobs after twenty years of intensive fishing.[5]

Concerning the Mediterranean, the stocks most in crisis are the tuna of the western area and the mullet and hake of the Tyrrhenian. Red tuna, in part because of the boom of the farms that I mentioned, is a cause of much concern. The International Commission for the Conservation of Atlantic Tunas (ICCAT), which controls the fishing of many species of tuna in all the seas, sets quotas of fishable tuna for any member country; but it cannot be effective against ship owners who fly the flags of nonmember countries and therefore, not being part of the commission, manage to evade the quotas. These tuna pirates thus overexploit the precious species, well aware of its high economic value, especially on the Japanese markets, and are not interested in the fact that the rate of its repopulation is diminishing.[6]

In Italy there is the additional risk that in a few years anchovies, cod, red mullet, swordfish, rayform fish, and sardines will become rare delicacies, like caviar.[7] The commercial species that are in sharp decline in the Italian sea, because of unregulated fishing, have been cataloged by the WWF and, to better explain the situation, we report the findings that this association has elaborated, underlining the causes that put these fish stocks at risk.

COMMERCIAL SPECIES AT RISK IN ITALY

Anchovy (*Engraulis encrasicolus*)
In the Middle and High Adriatic Seas there was a reduction of the stock in 1996. Presently 25 percent of the biomass is fished. It is recommended that this level of catch should not be exceeded and, in particular, that the minimal length of what can be fished be made equal to the length of fish in their first sexual maturity. Italy has strongly opposed this measure, which has not been adopted by the CGPM (Conférence Générale des Poids et Mesures; General Conference of Weights and Measures)

Hake (*Merluccius merluccius*)
In the northern Tyrrhenian and in the Ligurian Sea the stock has actually been reduced to 16 percent of its original size. It is recommended that a 15 percent reduction be imposed on the volume of the catch, that the areas of reproduction be seasonally closed, and that the zones of concentration of young cods be closed to fishing.

Mullet (*Mullus barbatus*)
It has been stated that in the southern Ligurian Sea and the northern Tyrrhenian, given the pressure fishing presently puts on the stock, there is a serious risk that it may collapse. It is recommended that the coastal zone be closed to fishing for as far as three miles, that artificial barriers against trawlers be installed, that there be a seasonal end to fishing in the postrecruitment period to avoid the capture of young specimen, that the "fishing effort" (i.e., the boat's capacity and motor power multiplied by the days passed at sea) be substantially reduced, and that the authorities impose strict controls over all these measures.

Sharks and Rayform Fish (*Chondrichthyes elasmobranchii*)
In the northern Tyrrhenian there has been a noticeable decline in their capture over the last fifty years. Some species have disappeared (e.g., *Rhinobatidae*). The *Squalus blainville* in the northern Tyrrhenian was once very commonly caught; presently, however, the population is in strong decline because of trawling.

A high level of accessory catching by the longline fleet is also presently reported. In the Ionian Sea, between 1978 and 1981 one specimen of this kind of fish was caught for every 1.6 specimen of swordfish.

Swordfish (*Xiphias gladius*)
In the central and southern Tyrrhenian, the longline fleet has caught

great quantities of swordfish that are not mature and undersized. The average weight, for example, is between 12 and 17.5 kilos. The albacore longline fleet catches, moreover, extremely small young swordfish, weighing less than 3 kilos. In the Gulf of Taranto and along the Sicilian coast, the albacore longline fleet has caught a number of young swordfish (less than one year old) equivalent to 53 percent of the total number of the specimen caught.

Sardines (*Sardina pilchardus*)
In the Gulf of Manfredonia, the fishing of the *rossetto* (a culinary delicacy consisting of another type of fish of very small dimensions in its adult stage) leads to the massive catching of immature sardines, amounting to 39 percent of the total catch.

Source: www.wwf.it/pesca/specierischio.asp.

If species are in danger because of the profit that can be drawn from them, those that are protected are certainly not safe either. The problem of accidental catches involving species that are safeguarded adds to the irreparable damages caused by excessive and illegal fishing. Still, according to the estimates of the WWF, about sixty thousand specimens of marine turtles are trapped in nets every year in the Mediterranean, while seven thousand dolphins and ten thousand porpoises are accidentally caught in the North Sea.[8] Industrial fishing boats use nets that are not very selective and can catch entire schools of fish, many of an inestimable value for the ecosystem and possibly protected species not in demand on the markets. These animals are considered bycatch and consequently are discarded and thrown back into the sea, dead or dying. Every year in India about 150,000 turtles drown, remaining trapped in very long nets without exit. This reptile, sacred to the people of the country because it represents one of the ten incarnations of Vishnu, the master of creation, thus risks extinction.[9]

If in Gandhi's land the victims are mainly the turtles, in the Eastern Pacific, another protected species, the dolphin, dies in the nets of seine fishing. In this area, tuna tend to swim under schools of dolphins, and to catch them fishermen tend not to worry about this generous friend of humans. On more than one occasion, in fact, this precious cetacean has saved inexperienced swimmers from drowning, and this is how Plutarch describes it:

> Of all the earth animals, many avoid men, and those that come close to them, like dogs, horses, and elephants, do so because they are fed by men. But to dolphins, unique among all animals, nature has given that gift that in vain the greatest philosophers seek: disinterested friendship. It has no need of man and nevertheless it is a generous friend of all, and it has helped many.[10]

Fortunately, in the last twenty years, there has been an attempt to reduce the collateral catching of dolphins and non-required species, thanks to the program Dolphin Safe of the Inter-American Tropical Tuna Commission (IATTC). The countries that are party to this program are Bolivia, Colombia, Costa Rica, Ecuador, El Salvador, the European Union, Guatemala, Honduras, Mexico, Nicaragua, Panama, Perú, the United States, Vanuatu, and Venezuela. These countries have agreed to utilize the system that Dolphin Safe prescribes, e.g., seine nets with thinner meshes in the part close to the edge, which allows the cetaceans to gain their freedom when the net reaches the surface. The boat that controls the net makes a reverse maneuver that gives free way to the animals. The crew, moreover, has the equipment necessary for immersion to free the animals that remain trapped.[11] This is a step ahead, then, in the reduction of accidental catches, but by itself it is not sufficient.

From many quarters it is demanded that the status quo of our seas be changed. However, much remains to be done to guarantee that fish stocks are available to future generations, who are entitled to them, and to whom we owe the respect that others have not had for us.

ARTISAN FISHING: A PATRIMONY AT RISK

The impoverishment of resources caused by this irrational exploitation creates hardship especially for coastal communities that draw their livelihood from artisan fishing and preserve an inestimable cultural patrimony tied to the blue world. When we speak of artisan fishing we refer to all the small fishing that is done with boats that are less than ten gross tons and twelve meters in length, are equipped with selective tools, and operate within twelve miles of the coast. The characterization of artisan fishing, then, is twofold, for it takes into consideration both the distance from the coast and the dimensions of the boats, which are usually rowing boats or boats with outboard or inboard engines of limited power. The tools used are usually fixed and very selective as far as kind and size (gillnets, longlines, traps, fishing lines, spears, etc.). But by artisan fishing we refer also to the traditional ways, much more poor as far as their equipment, used in the developing countries, but to some extent, surviving in regions all over the world.

Here is a description that makes us discover this kind of fishing, which is tightly rooted in the knowledge of the sea, in its resources, and in traditions and secrets that are transmitted from one generation to another:

> It is a simple, traditional way of fishing, hardly dangerous: a boat, one, two, three fishermen, rarely a boat too modern. The fisherman knows the sea in front of his port as the peasant knows the territory of his village. He knows all the places where it is logical to find the grouper, the gilthead, the soles, and even the turbot, the mullet, the gray mullet, and the hake, and he knows when offshore it is possible to catch sardines and anchovies (which will then serve as bait in the fishing of tuna). He exploits the sea like a peasant exploits his field. He never goes too far from the port or the cove of his village. If he raises his eyes he can see his house. And to go too far from the coast would mean to abandon the more fishy waters. He is an artisan and he fishes as it has always been done, with nets, traps, fishing boats, fishing lights, "yesterday a resinous torch, today an acetylene lamp or a lamp run by battery" that lights up at night: the source of light has changed but the principle remains the same.[12]

As a farmer does with their land, the fisherman as well cultivates as best as they can their sea. They know the species, the seasons in which they are fished, and traditional means of catching in extraordinary harmony with the natural environment. They know when to trust the sea, when to distrust it; they protect it, aware that it is the source of their livelihood. Small fishing also valorizes typical, local species and preserves activities rooted in the traditions of fishing villages. The social function of this fisherman, then, is crucial not only for food provisioning but also for the safeguarding of marine biodiversity.

In Italy, with more than eight thousand kilometers of coastline, artisan fishing can be described through the following figures:

Boats: 13,600

Annual production: 69,000 tons

Turnover: 415 million euro

Gross average price of the product landed : 6.01 euro per kilo

(Source: www.wwf. it/ambiente/dossier/Pesca%20in%20Italia.doc)

This fishing catches fish for the internal market, and many species are sold directly at local markets or to restaurants. Our fishermen, tied to their origins, keep alive, still today, ancient traditions. In Favignana, Italy, for instance, between April and May, they still practice the traditional way of fishing tuna.[13] On the island, where one of the last *tonnare* (tuna traps) survives, a reminder of the eight that could be found there until the end of the nineteenth century, the community gathers in a sacral group, a sort of tribe, to catch the precious fish. The *mattanza* (great slaughter), in fact, is not only a form of fishing; it has all the elements of a ritual. It is a practice tied to the festive time of the holy week, constrained in a space subdivided by instruments reserved for this use and carefully conserved, in a proper place, when the ritual is over. It is transmitted with precise rules, and above all it is isolated in a specific place, as it is typical of a ritual action where the direct outcome is death. The nets are lowered to the bottom of the sea, sustained by two lines of boats, and they form a corridor through which the tuna are forced to go. Until the last minute, when the net is closed, every gesture is regulated and directed by the *rais*, who has a refined knowledge of the whole.[14] His decisions, until recent times, were unwritten laws for fishermen and the entire community. "One became *rais* after a long career lasting the entire life, [working] in the *tonnara* from the age of nine to death, passing from the job of *foratico* to that of *arringatore*, to skipper, to vice *rais*."[15] Of this way of fishing, now reduced to very few instances, we have a recollection dating back to antiquity; it is from Aeschylus who, describing the battle of Salamis, refers to the defeat of the Persians using the image of the *tonnara*:

> The sea disappears under a heap of bleeding bodies. The Greeks rage against the Persians as if they were tuna caught by the net, breaking their backs with stumps of oars and fragments of wrecks.[16]

Among the populations of the Mediterranean, in fact, tuna has always been a source of survival. Not accidentally, in the vicinity of the small harbor of Favignana there is a kiosk dedicated to the Madonna representing her holding a tuna in her arms, a testimony to an ancient worship, proving the importance of the species that was assimilated even to Christ, who sacrifices for the good of humanity.[17]

Leaving the beautiful island and the rather bloody practice, we arrive in Sardinia, where fishermen preserve an ancient craft: the production of the roe. We refer to the mullet roe, which are the eggs that are extracted from the fish between August and October and then are salted and dried. The roe becomes so substantial that it can be grated and cut in slices. The phases of the fishing unfold here too according to ancient techniques. The fish, which always swims against the current, is channeled into a sort of funnel at the end of which it is captured, some times even by hand. In Tortoli, in the province of Nuoro, about fifty fishermen joined in a cooperative have been trying to protect the product by fighting for some time to obtain the protected designation of origin (in Italian, the *Denominazione d'origine protetta*).[18] Moving north, precisely to Monterosso in Liguria, there is still someone who keeps alive the art of night fishing for anchovies, *u pan du ma*, the bread of the sea, as the local elderly call them. Thirty years ago fishing was the principal activity of the place. Today, instead, only two fishing boats remain that fish with the *lampara* (fishing light), returning to shore at four or five in the morning.

Difficulties are not lacking, and in this locality too, people are trying to obtain a distinctive mark acknowledging the quality of the anchovies and allowing the salting to be done locally, which would guarantee a future to the now small fishermen community. These are only some aspects of the many realities that enable better survival along the coasts of the Italian peninsula, amid thousands of difficulties.

All the problems that we have discussed in the previous chapter, added to the aging of the fishermen, constitute a serious impediment to the possibility of prolonging the viability of small fishing.[19] But the loss of jobs in the fishing sector is a sad reality; in a recent ten year period, about sixty thousand European fishermen lost their employment.[20] To protect this patrimony, in November 2002, the General Association of Italian Fishing Cooperatives (AGCI) and the WWF signed an agreement for a plan of action for sustainable fishing intended to safeguard fish and fishermen. It is a set of concrete actions that see fishermen and environmentalists jointly engaged in safeguarding marine fauna and fish resources. Six areas of intervention have been identified:

1. Management of the protected marine areas with local operators from the fishing sector
2. Regulation of sport fishing
3. Initiatives to extend the application of the FAO code of behavior for sustainable fishing of 1995
4. Struggle against illegal fishing
5. Struggle against marine pollution
6. Expansion of tourism fishing activities and environmental education[21]

The agreement is undoubtedly something important for the preservation of our sea and the sustainable activities associated with it, but we are far from truly considering this activity like an art.

FISHING AND VALLEY CULTURE OF OTHER TIMES: A LOOK AT THE VENICE LAGOON

In order to think of fishing truly as an art it is sufficient to go back in time and arrive at the republic of Venice when fishing and valley culture were considered extremely important for the inhabitants of the estuary, so much so that the law and regulations of the republic subjected them to strict controls. Fishing was subject to the supervision of the *Giustizia* (Justice), a *magistratura*, or judiciary office instituted apparently in 1173 with the approval of the Ration Law (*Legge Annonaria*), which had the task of controlling the 123 arts that existed in the Doge's land. Vigilance was necessary because fishing and its products were the main source of food for the populations of the lagoon. It became clear, in fact, that in order to guarantee everybody the fish patrimony, its trade had to be disciplined. Nets in particular were made the object of special attention: they had to be marked by the Office of Justice and had to have the same length and meshes as the model deposited with the superintendents of this office. The fishing of new fish was especially protected, because according to *magistratura* records, "if the fish is allowed to grow and is fished in the appropriate seasons, it brings a joyous prosperity to universal benefit with the happiness of bountiful catches, a true remarkable advantage for the fishermen themselves." Also, the allocation of the places in the fishing of the new fish was done by drawing lots. At the beginning of March an urn was filled with the names of all the valley people and another with those of the fishermen of the swamps. The fishermen had to swear on the rules issued by the *Giustizia*. We report some here:

THE FISHERMEN CAPITULARY
In the name of our Lord Jesus Christ, so be it.

In the year of the Lord 1227, in the month of October, we officers appointed to exercise the justice, issue the following disposition concerning fishermen and fish sellers, so that it is inviolably observed by them, and we make everybody swear on the rules hereby listed:

1. I swear on God's Holy Gospel that from this moment to the next feast of Saint Michael I will not buy fish or birds of any quality to sell them back on land, and if I learn of someone who does the opposite of what has been said I will denounce him as soon as possible to the officers of the Justice who are or will be in charge. In the same way I will not buy either fish or birds in Rialto or in San Marco Square to sell them back on land and I will denounce to the officers of the justice as soon as possible all those who to my knowledge

have bought fish or birds to sell them back, as it is stated above. I will pay attention to all these things and will observe them, taking into account the prescriptions and rules that the Lord Doge will wish to add or eliminate, acting with the majority of his Council and the officers of the Justice who are or will be in charge.

2. Moreover, all the fish and birds that will be given to me to sell, I will sell them legally and I will give the money that I receive to the buyer/seller or buyers/sellers from whom I will have received the fish or birds, or will have had them or made them receive, except for the tenth part that it is legitimate for me to keep for the sale.

3. It is, moreover, established by our Lord Doge, his Council, and his officers of the Justice that whoever violates this prescription will have to pay a fine of 30 *libbre* and 12 *soldi* and that afterward he will not dare to continue in the same trade and if he will continue he will have to pay the fine mentioned every time he is found out.

4. In the same way, nobody can sell fish except those that are found in Rialto or San Marco.

5. We also want and order that every fish seller should not sell fish or birds that are spoiled.[22]

The chapter proceeds up to the disposition XXXVIII, which establishes, in addition to all the regulations for the buying and selling of fish, an apprenticeship of at least one year in order to become part of the art. Fishermen belonged in fact to a community, which was organized as a brotherhood (*La Fraglia*) with an elected chief called *gastaldo* and its own code of laws, called *Matricola* or *Mariegola*. The gastaldo was helped in his activities by twelve councilors, who were called presidents. The procedure that brought the election of the gastaldo after the death of the one in charge was solemn and took place in the church; the winner was made to leave the sacristy and was led to the main altar, where he pronounced a loyalty oath.[23]

This far-gone world survives in part in ethnographic museums, where we can go back, thanks to their collections, to the old trades. Particularly interesting, from this viewpoint, is the Museum of the Territory of the Valleys and Lagoon of Venice in the WWF Oasis of the Averto Valley in Lugo, in Campagna Lupia. In the museum are exhibited various artifacts and various boats for fishing central to the activities in the lagoon: spears for eels, sturgeons, and frogs; various kinds of nets; different kinds of containers for fish; tools for the making and mending of nets; various types of forks and bailers for boats and different types of rakes for hand fishing; different kinds of boats built specifically for fishing in the lagoon or marine fishing.

All these objects evoke the wisdom of a handicraft typical of other times: the *squerariòl* who built boats; the *remér* who took care of the oars and their support points; the *redaròlo* who once made the nets by hand, thanks also to the contribution of the *corder* who provided the proper ropes; the *favaro* who took care of the spears; the peddler *pescaòr* (fisherman) who would go to sell the catch in the hinterland; the *pessenovelante* who would fish the fry for the activity of the valley culture. This last activity required, moreover, a highly professional worker: the *capo vàe* (chief of the valley) who managed fish farming in the valleys. These were all professional figures who, in our time, if they have not completely disappeared, are certainly in great decline. To give an idea: the depositories of the art of the *remér* are less than ten.

Intensive fish farming, with its environmental impact, is putting into difficulty the classic extensive and sustainable valley culture and the vast range of knowledge that was connected with it. Until the first half of the twentieth century the vitality of this sector was reflected in the employment, besides the *capo vàe*, of many people engaged in the management of the *lavorieri*, which were structures made of fences, parapets, and shelters made of poles and branches intertwined to form a V, with the top placed toward the sea, through which the fish entered and could hardly get out. The ability of the valley chief resided in his wisely managing the waters of the valleys, regulating their salinity to make the different species of fish arrive at the *lavorieri*. Under him, in order of importance, came the fishing master, the huntsmen's chief, and the *omeni de vàlle* (men of the valley). Among them there were many connected professionals:

- the *chiusante*, whose task was to watch different types of *lavorieri* during fishing at the time of the *fraìma* (the migration of adult fish toward the sea);
- the *fangàro*, whose task was to constantly control the internal banks of the valleys; in some of these, temporary workers were employed who were called *furlani*, though they did not necessarily come from Friuli;
- the *barcàro da valle*, who transported the fish from the valley to the fishmonger shops on the appropriate boats (*batèe, caorline*, and *topi*);
- the *guardian de vàe*: the watchmen, often armed, who had to make sure that the fish were not stolen.[24]

Beside these figures, there were the valley's fishermen:

- the *reànti*, who dropped the nets at the time of the *fraìma* to catch the fish that were going out;
- the *cogolànti* in charge of preparing the *cogòlli* during Lent or the summer for the fishing of the *bisàtti* and the *gò* present in the valley;
- the *fossànti*, who caught the eels that had not migrated into the *lavoriero* and could be fatal for the planting of the new fish;
- the *trattariòli*, who had the task of fishing the fish that had not felt the instinctive call that would lead them to the *lavoriero*;
- the *ostregànti*, who collected, instead, the oysters that were often abundant and carpeted the bottoms of the valley;
- the *vendaòri*, who made agreements with the people and merchants of the valleys for the sale of the fish.[25]

In the pictures that follow we can see the *lavoriero* located in the oasis of the Averto Valley. It is good to imagine it as it once was, populated by all the protagonists we have discussed. Presently, however, the productivity of the Adriatic fishing valleys is approximately 70 to 150 kilos of precious fish (eels, basses, giltheads, mullets) per hectare, in contrast to the last century, when production registered higher average quantities of about 120 kilos per hectare.[26] Yet the dominant model of large-scale industrial fishing is destroying not only resources but also crafts and knowledges that we should instead take into consideration in order to reorient our approach to the sea in ways that can safeguard the wonderful wealth it offers.

NOTES

1. C. Merchant, *The Death of Nature: Women, Ecology and the Scientific Revolution* (New York: HarperCollins Publishers, 1980).
2. "*Situation mondiale des pêches de l'aquaculture*," in *Rapporto SOFIA 2002: La situation mondiale des pêches et de l'aquacultur*, in www.fao.org./docrep/005/y7300f/y7300f01.pdf.
3. *Squali, predatori a rischio*, November 11, 2002, available at: www.wwf.it/news11112002_9847.asp.
4. This evaluation is taken from *Libro Verde dell' UE: Relazione sullo stato ed evoluzione delle risorse ittiche*, available at: www.europa.eu.int/comm/fisheries/greenpaper/green/volume2ab_it.pdf.
5. *Merluzzo a rischio nei mari del Nord*, October 25, 2002, available at: www.wwf.it/news/25102002_330.asp.
6. *Ecologia e allevamento del tonno*, s.d., available at: www.marevivo.it/tonno3.html.
7. *Merluzzi e sardine, anche i pesci piangono*, April 2, 2002, available at: www.wwf.it/news/242002_6250.asp.
8. *Merluzzi e sardine, anche i pesci piangono*, April 2, 2002, available at: www.wwf.it/news/242002_6250.asp.
9. V. Shiva, *Stolen Harvest: The Hijacking of the Global Food Supply* (Cambridge, MA: South End Press, 2000).
10. *Mitici delfini*, s.d, available at: www.marevivo.it/arche.html.
11. *Tonno, la pesca sostenibile che salva i delfini*, May 9, 2003, available at: www.wwf.it/news/952003_440.asp; *Il sistema "Dolphin Safe" e le catture accidentali*, May 9, 2003, available at: www.wwf.it/news/952003_9840.asp.
12. F. Braudel, Il Mediterraneo, vol. 2 of La Méditerranée et le Monde Méditerranéen a l'époque de Philippe II, 3 vols. (Milano: Bompiani, 1987), 34. Originally appeared in 1949; revised several times).
13. In Sicily, of the eighty *tonnare* (tuna traps) that one could find until two centuries ago, only two have remained: the ones in Favignana and Bonagia. The article "Fuga dal Mediterraneo I tonni sono scomparsi" by A. Bolzoni in *La Repubblica* on May 9, 2003 highlights this worrisome development. There was no *mattanza* (great slaughter) in Sicily in the spring of 2003; the tuna in fact did not arrive at their death chambers. From the Atlantic through the Straits of Gibraltar to the Mare Nostrum, tuna arrived in great numbers for thousands of years, but that year they decided to go another way. The hypotheses made are: pollution and the passion of Japanese consumers for *Thunnus thynnus*. All the fishermen devoted to this type of fishing curse the Japanese "flying slaughterhouses" that dredge our sea chasing big and small tuna with sonar, and then work them onboard. The tuna is decapitated,

eviscerated, and frozen at less than 50 degrees Celsius to arrive frozen in the ports of the Rising Sun.

14. This terminology, still in use today, derives from the Arabs who, according to tradition, developed the systems of capture characteristic of the *mattanza*. The chief of the fishing expedition is in fact called *rais*, which in Islam is the title of sea captains. The *foratico* instead is the one who perforates the tufa to settle the nets, while the *arringatore* is the one who with hooks pulls the tuna over the edge inside the boat.

15. *Ecologia e allevamento del tonno*, s.d., in : www.marevivo.it/tonno3.html.

16. F. Braudel, *Il Mediterraneo*, 37.

17. E. Coppola Amabile, *E la Madonna cullava un puccolo tonno*, s.d., in : www.marevivo.it/arche2.html.

18. It is the cooperative of the Tortoli fishermen in a locality called Peschiera San Giovanni.

19. The average age of the operators is progressively increasing due to the lack of generational turnover. Young people are not very interested in fishing and often end the family tradition. Compare to: www.wwf.it/pesca/italia.asp. This fact is underlined also in the *Report SOFIA 2002*. The active population engaged in fishing is aging. In 2000, to give an idea, about 32 percent of the male fishermen working in Japan were older than sixty. Only 8 percent were under twenty-five. *Situation mondiale des pêches*, in *Rapporto SOFIA 2002*, available at: www.fao.org./docrep/005/y7300f/y7300f01.pdf.

20. *Il Wwf da Ancona lancia un appello per la pesca sostenibile*, March 25, 2002, available at: www.wwf.it/news/ 2532002_6250.asp.

21. *Pescatori e Wwf uniti per salvare il mare*, November 13, 2002, available at: www.wwf.it/news/13112002_4327.asp. The main points of the WWF AGCI agreement are found in the article *Accordo Wwf-AGCI Pesca, I punti principali*, November 13, 2002, available at: www.wwf.it/news/13112002.7987.asp.

22. Administration of the Province of Venezia, ed. *La pesca nella laguna di Venezia* (Venezia: Albrizzi, 1981, 14–17.

23. Cultural Association “El Fughero” ed., *La pesca in mare. Metodi, tecniche, esperienze di vita* (Venezia: Salvano, 1989), 10.

24. L. Antonini, A. Marcato, and G. Rallo, *Gli antichi mestieri delle valli* (Mestre: Museo del Territorio delle Valli e Laguna di Venezia, 2002), 8.

25. Ibid.

26. Ibid.

FOUR

THE FISHERMEN MOVEMENT

Mariarosa Dalla Costa

COASTAL COMMUNITIES AGAINST POVERTY

While in the Global North there is a growing awareness of the significant problems that the impoverishment of the seas generates, we must turn our attention to the South to fully understand the dramatic consequences of the predatory strategies that are typical of the fishing industry and fish farming. In this region the economies of the coastal communities had generally ensured their subsistence, thanks to traditional systems of fishing and fish farming, combined with traditional systems of agriculture. What has made it impossible for these communities to survive has been the advent of large-scale mechanized fishing and industrial aquaculture, which has led to a drastic reduction or the disappearance of the resources that had been the basis of their economies.

As a consequence, many artisan fishermen as well as many farmers have fallen into a condition of persistent and growing pauperization, forming in large part the 1.2 billion people who live in extreme poverty, on less than one dollar a day.[1] The excessive catches taken by the fishing boats of the rich countries deplete the waters of the South. Klaus Topfer, executive director of the United Nations Environmental Program (UNEP) has emphasized this.

> In many parts of the world fish stocks are suffering because an excessive number of fishing boats, benefiting from many financial subsidies, are drastically reducing the number of fish. Some developing countries with good fish stocks have stipu-

> lated fishing agreements with foreign countries in the hope of boosting the influx of foreign currency with which to pay their debts and stimulate economic growth. But our research indicates that, unless rigorous protection mechanisms are activated, this can be a dangerous mistake.[2]

European fleets are at home in the African seas, with often pernicious consequences for the local populations. Numerous agreements have been stipulated between the European Union and the countries of Africa, the Caribbean Islands, and the Pacific. Emblematic is the one signed with Mauritania on August 2, 2001, allowing for access to the country's waters in exchange for €430 million. After months of intense negotiations and the signing of the agreement, the local populations have many reasons to fear.[3] West Africa, after many years of European fishing, has lost half of its ground fish stock, a category that includes some of the species that are most valued from the commercial viewpoint. Daniel Pauly, a world authority on the study of the global exploitation of fish resources, declared at a conference organized by WWF International in Dakar (Senegal) in 2002 that

> because of the nonsustainable exploitation of fish resources by foreign fleets, West African ecosystems are now as impoverished as those of the North Atlantic, but the consequences on development and food security in their case are extremely serious, far worse than those in Europe or North America.[4]

To this landscape we must add fish farming factories; these too, proposed with the humanitarian intent of increasing the food supply, have had a destructive impact on coastal areas, causing malnutrition and the exhaustion of marine resources. While more than 75 percent of the fish consumed today by human beings still comes from free species living in natural ecosystems, aquaculture is the fastest growing system of global fish production, and in tropical countries at the top of the list is shrimp. Across the world, more than half the consumption of salmon and shrimp is industrially produced instead of being fished in the sea. In any case, the industrial fishing of shrimp done with trawlers has produced an enormous quantity of discarded fish. Indeed, if the global annual waste produced by commercial fishing companies can be prudently estimated as twenty-seven million tons, equivalent in weight to more than a third of what is globally fished commercially, the waste caused by the motorboats catching shrimp and prawns is the largest of all, for annually sixteen million tons of dead or dying fish are thrown back into the sea. In some fishing zones up to fifteen million tons of fish are discarded for every ton of shrimp that is caught.[5]

In the same way as the intensive monocultures of the "green revolution" function in agriculture, similarly the tubs of what Vandana Shiva calls the "blue revolution" operate in the fish sector, monopolizing the very rich humid zones of many developing countries.[6] While the high-returns strategy pursued by the agricultural revolution eliminated the legumes and oil seeds that are essential for the diet of the populations and the fertility of the soil in India, and while many African countries depend on the revenues drawn from one monoculture crop for export, the intensive fish farming factories have swept away entire mangrove forests in Ecuador, Bangladesh, China, the Philippines, Honduras, India, Indonesia, Mexico, Thailand, and Vietnam (the main world producers).[7] The consequence of the development of these farms, catering to the production of crustaceans for Western tables, has been highly destructive in the South of the world, and they do not contribute to the diet of the local residents but actually worsen their nutritional and living standards. The terrains that the multinationals acquire are dug up to accommodate tubs two meters deep and covering an area of about one hectare. Just one of these tubs in India guarantees a revenue of one million rupees annually.[8] The basins, later filled with salted water, are ready to host millions of shrimp at the expense of the land, the water, the biodiversity, and therefore the wealth of the coastal populations. The loss of the mangroves, nurseries to many fish species, takes bread away from small-scale local fishing and exposes the coasts to a more rapid erosion. The different species of fish move to deeper and more quiet waters, making the work of artisan fishermen harder, as it takes them more time to catch the same quantity of fish needed for their livelihood. Many fish, moreover, die, because the water is polluted by the illegal drugs used to protect the shrimp that are packed in the tubs—water later flushed into the sea and the neighboring areas. As their natural protection is lost, the coasts are more exposed to natural catastrophes. Fish farms and factories have turned the land into a scarce resource and reduced the quantity of potable water, due to the salinization of the subterranean aquifers.[9] These farms, located between the coast and the hinterland, require, in fact, constant inputs of salted water and mechanical motion to oxygenate the crustaceans. The salt penetrates the terrain and expands through the aquifers, causing a desertification process in the neighboring areas for square kilometers. The lack of cultivable land, the scarcity of water, the consequent death of animals, and the increasing destruction of the ecosystems that guaranteed subsistence have a disastrous social impact.

Many local fishermen and farmers' communities are displaced due to land expropriation and forced to abandon the areas that for centuries had ensured their livelihoods.[10] In Ecuador, for example, one hectare of mangrove forest guarantees food and subsistence to ten families, while a shrimp industry of 110 hectares gives work to only six people. In Sri Lanka, twenty thousand fishermen of the Puttlam lagoon have been forced to leave.[11]

In India women are forced to work from four to six hours a day to collect water and wood to burn, which have become scarce because of too many aquaculture plants.[12] It is above all women and children who work in the "blue factories" from eight to ten hours a day in terrible hygienic-sanitary conditions. The children of the fishermen and the farmers, due to their families' poverty, are forced to leave school and work in the tubs in conditions near slavery for an average salary of thirty-five dollars a month. The local NGOs denounce, moreover, many cases of rape perpetrated against female workers.[13]

In the areas near the farms skin diseases and, above all, endemic diseases such as diarrhea, substantially increase, worsening living conditions and striking above all the weakest: the elderly, children, and women.

Also, the processing of the shrimp creates situations in which very small children work in horrible conditions. An example is Machar Colony, a colossal slum that has grown substantially in recent years near the harbor of Karachi, Pakistan, in the "fisheries area." This is a sort of no-man's-land populated by over 700,000 people—Bengalese, Burmese, and above all Afghani—almost all clandestine, who have fled wars and coups d'état. The most substantial business, controlled by the country's mafia bosses, is the processing of these crustaceans, which relies on the intensive exploitation of children. Crouching in long lines on wet and stinking floors, they shell mountains of shrimp for twelve hours a day, under the obsessive surveillance of guards, who prevent them from talking with each other and wasting time. Every small transgression is met with corporal punishment. The pay is set on the basis of the number of fish baskets they fill. Those who in a day manage to process fifteen kilos can earn two dollars. Because of the position in which they must work, and because they must keep their hands in salted water in which ice is mixed with the shrimp, these children are destined to develop arthritis in their fingers and spine problems. Migrant parents often deliver their children's wages to pay back those who helped arrange their clandestine trips to Pakistan. It is calculated that the mortality rate among children under five in Machar Colony is 250 per thousand, twice the average in Pakistan.[14]

The choice of industrial aquaculture over the traditional one leads to perverse results. Like the green revolution before it, industrial aquaculture was presented as a contribution to the solution of the problem of hunger in the world, having presumably the further merit of compensating for the lack of proteins in the diet of the poor and reducing the pressure on the sea at a time when resources were becoming scarce. But the litmus test for the gap between promises and reality is the fish feed. To catch the quantity of fish necessary to produce the feed for farmed shrimp it is necessary to increase the pressure on the sea, not reduce it. This industry, in fact, requires from four to six tons of feed per hectare. In 2000, the industrial fish produced in Asia was estimated to be about 5.7 million tons. The quantity of fish feed necessary to produce this amount is about 1.1 million tons of fish, which is derived from the equivalent of 5.5 million tons of wet fish—that is, nearly twice as much the total amount of what is fished in India. The catches are obtained with motorboats with seine nets, which notoriously lead to the exhaustion of marine resources. Moreover, only 17 percent of the fish feed is converted into biomass contributing to the growth of the shrimp. The rest, highly polluted by pesticides and antibiotics, is thrown back into the sea or flushed into the mangroves or the neighboring cultivated terrains. The blue revolution, then, like the green revolution, by consuming more than it produces, represents a false promise of abundance.[15]

The uniformity of the productive/destructive model characterizing the blue revolution obviously does not take into account the different traditions, the local knowledges, and the mutual relations tying human beings to the natural elements. Unlike the fish factories that are imposed on coastal populations, the traditional farming systems were sustainable socially and environmentally, and for centuries they were a guarantee of food security for fishermen and farmers.[16] As Shiva points out, these systems, in existence for five hundred years, are local systems that have a very light ecological impact and are as profitable as the industrial ones.[17] These systems made India the main producer of shrimp and guaranteed the food security of the farmers and fishermen of the coastal zones. The *bheri* system of aquaculture developed in the muddy areas of the tidal and malarial swamps of the Upper and Lower Sunderbans in West Bengal. *Bheri* is the name of the ponds that characterize it, which have a dimension variable from 2 to 267 hectares. There are two types of *bheri*. There is a seasonal one, which is in use from November to December and then left to dry in the sun until the following season. And there is a perennial one, which is found only in areas that have a high degree of salinity, where rice does not grow, and it can be destined therefore for farming fish and shrimp throughout the year. In Orissa we have, instead, the system of *gheri*. This is what ponds for aquaculture are called when located near estuaries, beaches, and lakes. They are enclosed by bamboo sticks tied with ropes, and nets are used to keep in the fish and shrimp. The tide pushes fish, shrimp, and other aquatic organisms into the nets, which they can no longer escape, and they feed on the food brought by the tide. When they are grown enough they are harvested. This traditional shrimp aquaculture method has been practiced for centuries in the low waters of Kerala. In the same fields, fish farming alternates with the cultivation of corn in the months of the monsoons, from July to October. After the corn is harvested, people grow rice that in turn is harvested before the tides flood the fields. For the cultivation of rice, the bottom of the field is raised to increase its exposure to the sun and to allow the superfluous water to drain toward the bottom. The internal waters increase the fertility of the soil with the nutrients and minerals they contain and transport. After the rice harvest, for the rest of the year, on the same terrain, shrimp and other fish are farmed. Salted water is allowed to enter the fields, but when the tide begins to retire, a tightly intertwined bamboo barrier is inserted in the opening to keep the young shrimp in. This simple maneuver is repeated at every tide throughout the whole period until the harvest, which is carried out by using sluice gates or launching nets or by hand. Another system is the *thappal*, which literally means "search." During the high tide the fishermen use their hands to catch shrimp, oysters, or other fish, which are then put in a container of salted water. The catching is facilitated at times by the immersion of a mat made of dry grass and balsamic plants intertwined with rice grains. The grains attract the shrimps that remain trapped in the mat. It is an ancient and well-articulated knowledge that the logic of profit has forced people to abandon and now is in danger of being lost. As Shiva again underlines, an exclusive focus on profit destroys the social conditions nourishing the intellectual diversity that is the basis of a people's creativity.[18] By not taking into account the complexity that is characteristic of nature, we endanger the very survival of human beings, who displace the environment and then are in turn displaced. But those who try to change the status quo that has been created in the name of money for the sake of a model capable of guaranteeing life risk their own lives. While the industrial vessels empty the sea, and while aquaculture destroys the humid areas, leaving to the indigenous people the landscapes we have described, fishermen and environmentalists who oppose this situation are threatened and often killed. In eleven countries, murders have been denounced in connection with the shrimp industry.[19] The destruction of the ecosystems also carries away the men tied to them.

FARMS THAT KILL

Industrial shrimp farms have given place to innumerable cases of violence and intimidation perpetrated against the local populations that decided to resist them, thus affirming their right to fish and cultivate, and above all, their rights to maintain access to the resources of the sea and the land on which they lived.

In 1988, in Honduras, the local population, with the help of environmentalists, founded the Committee for the Defense of the Flora and Fauna of the Gulf of Fonseca (CODDEFFAGOLF), a nongovernmental organization devoted to the defense of the environment and the protection of the local rural populations. In 1996, the committee obtained from the government a moratorium establishing that for one year there would be no new licensing of fish farming plants. This provision, however, despite the efforts of the environmentalist groups and the committee, did not prevent the construction of sixty new plants. The following year, three thousand people marched on Tegucigalpa, manifesting their opposition with slogans that well demonstrated their determination to stop the ongoing destruction. "No more shrimp industries," "Stop the wreckage of our communities," " Stop the destruction," they shouted, calling for a convention to protect the gulf. But the committee was opposed by the aquaculturers lobby that in the end prevailed, preventing the moratorium from having any effect. In the conflicts between the committee and the aquaculturers, twelve people died. In October 1997, two local fishermen, Israel Ortiz Avila (thirty years old) and Marin Seledonio Peraltra Alvarado (twenty-seven years old) were tortured and killed in the natural reservation of "La Iguana" because they did not want to leave their land. Some years before some other young men had been killed in the canals connecting the shrimp "lakes" to the gulf. The industrialists justified their actions by claiming that the fishermen had tried to steal the crustaceans from the plants, and only few spoke about these events. The committee, directed by Jorge Varela, himself a victim of threats and intimidation, after the murders wrote a letter to the government:

> Since the shrimp firms have taken the area into concession, local fishermen can no longer move freely through the estuaries and mangroves where before they were fishing for their livelihood. They cannot fish not only in the areas given in concession but even in the nearby areas. With the complicity of our system of government we have given up our patrimony to a few Hondurans and foreigners, depriving thousands of their means of subsistence, transforming the blood of our people into an aperitif and our natural resources into a dessert for the palates of foreigners.[20]

These are sentiments shared by all those who see the sources of their livelihood taken away without this crime having any resonance in the international community.

In 1999, in the state of Nayarit, Mexico, on the unpolluted beaches of San Blas, the creation of enormous fish factories produced a true ecological catastrophe in the natural park of Boca Cegada, which protects the last forests of mangroves of the Mexican Pacific. After cutting and burning two thousand hectares of forests, after blocking the natural canals, diverting and draining a good part of the brackish waters of the lagoons of Los Pajaros and El Zapato, the firm Granjas Aquanova SA poured into the waters of the lagoons industrial quantities of antibiotics, with irremediable consequences for the neighboring flora and fauna. Following this ecocide, some groups of ecologists and citizens sent denunciations to the Convention on Wetlands of International Importance, called the Ramsar Convention. The government responded by sending a commission to the area and stipulating an agreement with the firm for the construction of twenty-eight more plants along the country's coasts.[21]

In Brazil, in April 2002, Sebastiâo Marquez de Souza was killed by two individuals connected with the shrimp industry. A unionist active in the struggle for the protection of the Brazilian mangrove forests, he was among the many victims of this type of farm.[22]

In Bangladesh, landless farmers are the main victims of hunger and malnutrition. Access to land is a fundamental condition for their survival. Lands recently emerged on the rivers' banks or already owned by big landowners (*zamin-*

dars), who have had control of vast land properties since colonial times, were declared "public lands" (*khas lands*) by the Supreme Court. As such they could be distributed to landless people according to what was established by the agrarian reform, and the process had even taken off. But the desire to appropriate lands particularly fit for the very profitable shrimp industry led to bloody clashes among the interested economic groups and the local populations. One of the most violent cases of eviction took place in the district of Debhata and Kaliganj, in the southwest of the coastal region, on a terrain of about three thousand acres farmed by one thousand two hundred families from the villages of Baburabad, Kaminibasu, Katmahal, Bhanganmari, Dhebukhali, Kalabria, Norarcharkuni, and Jhaiamari. The land, once the property of two big landowning families, almost fifty years ago was declared "public land," although the owners continued to enjoy its possession through a new business. They tried again to regain the land, but in 1982 the Supreme Court reconfirmed its status as public land and its distribution among the landless peasants began. At the end of 1970s, thousands of hectares of cropland in the same area had been appropriated because of their high potential for the shrimp industry, due to the great influx of salted natural water coming from the rivers, which arrived at the nearby region of Sundarban. This increased the pressure on that land by those who had economic interests. The expropriation took place with the complicity of the local authorities. In May 1999, the police forces, helped by criminal bands paid by the power elite, carried out brutal evictions, including the destruction of two hundred houses. Fifty people, among them fifteen women, were wounded in Jhaiamari. In July of the same year, in response to the protest of those who did not want to leave the area of Baburabad, the police began shooting, killing a woman and two children and wounding another twenty-six people. During these years, in the coastal areas, 150 people died in "incidents" directly connected with the expansion of the shrimp factories. At times the land was polluted on purpose to make agriculture and cattle rearing impossible and lower its price, so that it could be bought more easily.[23]

In Malaysia, since 1991 more than half of twenty-five hundred acres of mangrove forest has been destroyed in Kuala Muda, an area in the state of Kedah, to make space for the shrimp industry. In the years following the cutting of the precious plants the earnings of small local fishermen dropped from thirty ringgit a day to five ringgit. The economic loss has been compounded by the fear that their villages may be destroyed by more devastating climatic events, as they are no longer protected by the forests. The contested project to open industrial shrimp farms in Kerpan, also in the state of Kedah, is another example of how the development drive supported by the government comes into conflict with the traditions and practices of the local communities, which can guarantee people's subsistence, something that the new economic choices do not allow. In 1993 about a thousand acres of land, formerly used for rice cultivation, were acquired by the state at the expense of eight hundred proprietors and about a hundred families who worked there. The lands were given to the Samak Aquaculture project, a joint venture controlled by a company from Saudi Arabia (60 percent) and by the state (40 percent), without any of the compensations promised to the former owners of the terrains and with losses also for the state of Kedah. Many were forced to sign the pro-Samak agreements without reading them, others opposed the project and organized an action committee that filed two lawsuits to challenge the expropriation. To the legal action they added civil disobedience, blocking the bulldozers. But the protest ended with thirty-three arrests.[24]

In Thailand, the responsibility for the destruction of half of the forests falls significantly on industrial crustacean farming, and as such the local populations have suffered serious losses. While the profits of the aquaculturers deliver the country to the interests of the multinationals, and the nutritive properties of the shrimp flee to foreign countries, small fishing is reduced to the bone. The Yadfon Association reports that Thai rice farmers have had to abandon their cultivations because they could not survive, due to the water infiltrations coming from the nearby aquacultures. In the summer of 1996, three thousand farmers without work protested in front of the office of the prime minister in Bangkok, denouncing the loss of their earnings.[25]

In 1999, in Indonesia—in southern Sumatra and in Lampung—hundreds of farmers were chased from their lands, and many were arrested to make space for destructive farms in conformity with the goal of the government, which wants to drastically increase fish production for export. The Protekan 2003 program aimed, in fact, to raise the revenues of the sector up to ten billion dollars within the following three years. Of these, seven billion were to come from shrimp factories, with a consequent expansion of the coastal area destined for the plants. Indonesia exports about 50,000 tons of shrimp a year, produced in farms that cover 360,000 hectares of terrain. The aim is to export 677,000 tons a years with the installation of farms in twenty-five new provinces, transforming another 320,000 hectares of land, almost twice the present figure.[26]

In the province of Aurora, Philippines, between the Sierra Madre and the Pacific Ocean are still exceptionally rich rainforests. It still has half of it—430 hectares as opposed to the 3 percent remaining in the rest of the country. In this very place, in 2000, both population and forest were threatened by a new expansion of the Diapitan Resource Development Corporation (DRDC) that, already in operation since the 1990s, compromised what remained of the forest. Beginning in 1997, the residents presented protests and petitions, but in vain. The whole environment has been damaged by the great quantities of industrial feed, antibiotics, bleaches, and other chemical substances being poured into it. According to what the residents of the villages of Masagana and Maligaya denounce, the groundwater aquifer has become salinized, fishermen find very little fish, harvesters of mollusks and shells suffer from skin irritations caused by the chemicals discharged: often waves of dead or deformed fish arrive, corals near the bank die. Ignoring the protests of the local population, the company plans to further expand its operations, eating away another piece of mangrove forest at Casiguran. For this reason a coordination of environmentalist and citizens' associations, the Aurora Support group, has been formed.[27]

In December 1996, in India, the Supreme Court ordered that before March 31, 1997, all the coastal farms had to be closed except for the traditional ones, including those that had been improved. Despite the court sentence, however, people had to continue to struggle and die for its implementation. While there should not have been any farms in the five hundred meter strip from the sea along the seven thousand kilometers of coastline, and in the thousand meters near the lakes of Chilika and Pulicat, the farms have not been removed.[28] The power and political influence of the multinationals have prevented the judgment of the court from being applied. In 1999, a demonstration against the illegal shrimp farms on Lake Chilika by the local fishermen's organization called Chilika Matsyajibi Mahasangh, led to the destruction of eleven illegal farms. In retaliation, the police burst into Sarano, a small village facing the lake, killing four people, among whom was a woman. Although the area wetlands are considered of international importance and consequently are protected by the Ramsar Convention, Tata Group, a powerful Indian multinational corporation, wrangled permission to install a number of farms there. The company was ultimately blocked, but its place was taken by a number of lesser-known investors operating in complicity with the local elite.[29] For decades in India people have been struggling against these types of plants, and the verdict of the court represented a victory won by defying the powerful lobby of the shrimp industry. However, the government seems to be ever more indifferent to this issue; thus, in order to have the court's decision implemented, a relentless effort must be made by environmentalists, local movements, and above all, as we will see later, the national and international fisher people movement.

The verdict by the Indian Supreme Court arrived after years of protests and mobilizations organized by various movements. First there was the constant actions—marches, fasts, nonviolent sabotages, legal suits—undertaken by the activists of the Gandhian movement Sarvodaya, which means "well-being for all" and is present above all in the villages of the southern state of Tamil Nadu. The struggle against the aquaculture companies, which boasted up to 140 lawyers, has been led by the movement of farmers, fishermen, and day laborers of the Grama Swaraj (movement for the autonomy of villages) and of LAFTI (Land for Tillers' Freedom).[30] At the leadership of the two movements are Jagannathan and his wife, Krishnammal, who, helped by an activist lawyer in New Delhi, M. C. Mehta, obtained the important success of a favorable verdict, proof of the unsustainability of the aquaculture farms. Jagannathan and Krishnammal have been working for more than forty years for the rights of the poor and against all types of violence.[31] They continue to be engaged in the effort to chase away these "environmental grasshoppers," and the "unaware shrimp," and to return the lands to the people who lost them. The first is the president of the Grama Swaraj, the second is the general secretary of LAFTI.

The objectives of the Grama Swaraj are a testament to the fact that the movement endures in the twenty-first century. In accordance with Gandhi's words, "The land belongs to God, that is to the community, the industries to the villages, and the power to the village." For Grama Swaraj, the political self-determination of the villages is the path indicated by God for peace in the world. The movement wishes to be the torchbearer of this peace, demolishing, without the use of violence, the three evils that plague it: the accumulation of wealth and the power-money nexus and capitalism; the concentration of land in the hands of a few and the power of the landowners; and the concentration of political power.

The efforts of Grama Swaraj, which organizes marches and demonstrations to call attention to the devastation of the resources perpetrated by the multinationals, are compounded by those of LAFTI. This last was founded by the couple mentioned above in 1981, on the basis of the Bhoodan (gift of the land) movement of Vinoba Bhave, "the purest disciple of Gandhi" who, since the 1950s, had been concerned with distributing land to day laborers.[32] LAFTI includes the residents of about five hundred villages and continues to engage in a struggle to promote village democracy, redistribute land, safeguard the environment, improve the conditions of women, and educate the poorest children in schools. During his life, Vinoba Bhave contributed to the redistribution of up to four million acres.

The following, which the perseverance and tenacity of the founders of LAFTI has generated, found a further confirmation on September 9, 2002. On that date, since the verdict of the Supreme Court was not being implemented, ten thousand people gathered in a large protest against fish farms, the main cause of the floods that in January had destroyed the entire rice crop in the Muthupettai, another area that has been deprived of the precious natural protective walls the mangroves provide.[33] Catastrophes like this continue, one after another, each with heavy consequences for the loss of human lives and resources; even as we write, it is evident that the destruction of the mangroves worsened the effects of the tsunamis of December 26, 2004, in Southeast Asia, leaving the territory with its front doors wide open to the eruption of the ocean.

OTHER SIGNIFICANT STRUGGLES

The movements and struggles in which coastal communities are engaged are articulated around a set of issues that—as we shall see later—keep multiplying. It was above all in the 1990s that the actions carried out by fishermen communities in India acquired a national dimension, in correspondence with the leap that occurred in the process of fishing and in industrial fish farming, in the context of the neoliberal political agenda that since 1991 has been imposed on this country. In the same year, the Indian government instituted the MPEDA (Marine Products Export Development Authority) to more strongly support aquaculture. Indeed, the agency has provided technical assistance and significant subsidies to this sector in the country.[34] If we keep in mind that 60 percent of this nation's one billion inhabitants live along the coasts, we can easily imagine what the impact has been of constant penetration into the territory and into the sea of fish farming and fishing methodologies that are so destructive to local resources. In areas where the population has always kept a friendly relation with the sea, this relation is now totally negated. In the north of Tamil Nadu, for example, the population involved in fishing, according to Goldy M. George, numbers thirty thousand people.[35] Traditionally it was not a fishing monoculture. But starting in the early 1990s, as very high profits were looming on the horizon with the advent of neoliberal politics and structural adjustment programs favoring a development primarily devoted to exports, big companies jumped into the aquaculture business, focusing on the United States's and English markets, and then generally the European market. Significant in its dramatic character, but not at all isolated, is the case of the village of Thiruvarur, again in the Tamil Nadu, where there are more than ten companies active in this sector. The free movement of capital invested here to produce for export has brought about the destruction of the resources on which people lived, the alteration of the ecosystem, and very heavy pollution. Hundreds of families had to leave the territory and pour into urban slums. Just from the village of Perunthottam in the district of Nagapattinam, as the author of the text just quoted again tells us, two hundred families had to abandon their habitations to move to the periphery of Chennai in search of means of subsistence. But it is not just people. Tens of thousands of varieties of birds that populated the lagoon near Muthupettai also had to leave because of the destruction of the mangrove forest caused by shrimp farms. It is a story that repeats itself wherever the tubs of industrial fish farming arrive: the destruction of life's resources; the destruction of the very possibility of survival for the inhabitants of these areas; the destruction of the infinite variety of species, animal and vegetable; the destruction of the environment; and, above all, the destruction of the knowledge and cultures, the practices and relations with the sea and the land, that formerly had enabled people to live.

As for fishing, in the very fish-rich tropical sea surrounding India, fish stocks are also at risk of becoming extinct if catches are not drastically reduced or a moratorium on fishing is not introduced. Species that constitute the primary source of food for the populations of the coast are at risk of disappearing. The fact that there was a reduction of catches worldwide in the nineties is in large part attributable to the fact that between 1970 and 1990 the dimensions of the fishing fleets increased twice as much as the rate of growth of the fish caught globally.[36] This very explosion in the number of fishing vessels, as Nandan, the ambassador of the Fiji Islands, underlined in 1997 at a session of the General Assembly of the United Nations concerning Agenda 21, has contributed to undermining the survival of fish-rich areas and the very vitality of the fishing industry.[37] In India, as elsewhere, coastal communities are more and more determined to engage in militant forms of struggle against the industrial fishing and industrial aquaculture that exhaust their nutritional resources. Four great fishermen strikes exploded in India between 1991 and 1997. But coastal communities have also organized hunger strikes, sit-ins and marches; blockades of highways, railroads and airports; and occupations of government offices and ports. This series of actions, coordinated by the National Fishworkers Forum (NFF) that has constituted itself in India since the 1970s, an organization to which we will later return, brought a regulation to the coastal zone (Coastal Regulation Zone, or CRZ) and the constitution of an agency that was to supervise its implementation. The determination of the residents of the coastal zones to defend their lives is clearly expressed by Pudi Kuppam of the village of Puttupettai who, as Goldy M. George again reports in the text quoted above, declared: "We live on this Earth. The sea is our mother. It takes care of us. If shrimp farming takes off we will be ruined, but we will destroy it. Wherever it is, we will eradicate it."

This situation was brought to the attention of the Global North with the initiative of the intercontinental caravan that was formed by various social movements from the South—farmers, fishermen, indigenous and tribal peoples whose possibility of subsisting were increasingly undermined by neoliberal politics. The caravan reached various localities in Europe between May 22 and June 31, 1999, providing an occasion for debate and consciousness-raising. Its objective was to bring to the industrialized West the viewpoint of the South.[38] It was to make known the system of exploitation and marginalization that the multinationals have imposed and the international institutions, starting with the World Trade Organization (WTO), have guaranteed.[39] The initiative came from the Karnataka Rajya Raitha Sangha (KRSS), also known as the Karnataka State Farmers' Association, a movement inspired by Gandhian philosophy that includes the participation of ten million farmers. The organization played a leading role in networking regional struggles through the Peoples' Global Action, an international coordination of movements against neoliberal globalization founded in Geneva in 1998.

The KRRS, born in 1980, includes small and middle-size farmers and landless peasants. It has a key role because it is not only concerned with the mobilization of the people of Karnataka but also involves at the national level other movements of the same type. It deals with a broad range of questions: resistance to the green revolution; consciousness-raising about the negative impact of modern biotechnologies in agriculture; opposition to neoliberal politics in trade; and the teaching of alternative agriculture based upon self-sufficiency, sustainability, and traditional methods of cultivation. Another prerogative of the KRRS is the organization of programs and meetings for women, valorizing their role in society. It thus proposes an alternative model of development, a model owned by the local communities who take charge of the decisions that concern them at a political, technological, and economic level. In order to maintain its autonomy, the association does not accept foreign funds nor contributions from political parties, the state, or the government. The money that it uses comes from the quotas paid by its members and from fundraising done during demonstrations.

Another important presence inside the caravan was the Bharat Kishan Union (BKU). Born at the end of the 1970s in the state of Tamil Nadu, the union supports the farmers who resist the green revolution. Within this movement, however, the idea of privileging self-sufficiency in the agricultural sector, freed from the development of neoliberal trade, is not shared by all. The union tries to adopt a political dialogue on this matter, while giving voice to farmers who do not accept fertilizers, biotechnologies, and all that distorts traditional methods of agriculture. The BKU participates in the Joint Action Forum of Indian People (JAFIP) against the WTO and its unpopular politics. Also a member of the caravan, the JAFIP was formed in 1998 precisely to demand that India withdraw from the WTO. This had been the largest forum of popular movements against the WTO and in the decade since then the movement had continuously expanded as it has included struggles from different sectors.

Also participating in the march was the Narmada Bachao Andolan, the movement to save the Narmada, born in the mid-1980s to fight against the Narmada Valley Development

Projects that planned the construction of numerous dams, thirty of them big, hundreds of average size, and thousands small, along the Narmada River, which traverses the states of Madhya Pradesh, Gujarat, and Maharashtra. The movement to save the Narmada represents a turning point, a new form of activism on the environmental front. It did not fight any longer just to obtain reparations for the damage done but put forward the population's affirmation of its right to decide about its environment and its system of life. Despite direct actions, such as blockades and hunger strikes, it had managed to block only a few sites. But the Andolan continues its efforts to this day to stop dams financed by the government and multinationals.

Alongside these movements was the National Fishworkers Forum (NFF), a coalition of farmers, fishermen, and environmentalists who wanted to bring their viewpoint to the North, building strong and concrete solidarity bonds with European civil society.[40] The caravan's objectives were above all informative, aiming to promote awareness of realities otherwise not well-known. These protagonists from the Global South came to Europe with this initiative, wishing to spur a broader process of aggregation and cooperation among organizations sharing the same principles. They wanted also to support nonviolent direct actions and forms of civil disobedience. Indeed, the caravan was to have a key role in the organization of the actions that took place in November 1999 in Seattle. In order to achieve these objectives the Indian movements' European tour was enriched by meetings, demonstrations, conferences, and nonviolent actions of civil disobedience that were held in front of the headquarters of multinational companies and political institutions and were to result in the contestation of the summit of the G8 in Cologne. The caravan initiative thus was one of the first seeds of the activities that, with the many others organized through People Global Action, would later oppose the decisions of WTO, demonstrate against the Millennium Round, and subsequently organize world and European social forums to frame the terms of a more equitable social project.

At the Seattle countersummit against the politics of the WTO, the fishermen's movement denounced the destructiveness of industrial aquaculture and large-scale trawling. The late Thomas Kocherry (1940–2014), an eloquent and eminent speaker from the fishermen's movement, won the Sophia Foundation prize in the same year, together with Herman Daly, for their joint focus on alternatives to globalization and denunciation of its pernicious effects. Father Kocherry was the president of the National Fishworkers Forum of India from 1982 to 1995; he was also the coordinator of the National Alliance of People's Movements (NAPM) in India from 1997, and from the same year was the coordinator of the World Forum of Fish Harvesters & Fish Workers. In his passionate speech on the occasion of the prize ceremony, he said:

> More than 100 million people in developing countries [in the South] are dependent on fisheries for their livelihoods. For us fishing is a way of life, not just a source of income. The sea is our mother. Traditionally, small-scale or artisan fishers have provided fish for local consumption; but as fish become scarce and its value increases, it enters the global market and becomes unaffordable for common people. In the process we are displaced and the multinational corporations take over completely. Most governments, particularly of the North, are trying to prop up an unsustainable fishery. According to the FAO, every year governments worldwide pay US$116 billion to catch the equivalent of $70 billion worth of fish. Developed nations, which have overfished their own waters, have headed into the waters of the developing nations. The European Union has around 40 percent more vessels than necessary to catch fish on a sustainable basis.
>
> Volatile "fish wars" are commonplace. There are more than one million large industrial fleets in the world. They have depleted all the oceans in the world. They have become a threat to the one hundred million fisher people in the world. Further, these have organic links with the coastal mono shrimp culture. Fresh fish caught by the industrial vessels are converted into fish meal for the production of shrimp. Ten thousand tons of fish that would have been available for common people are converted into fish meal to produce one thousand tons of shrimp that only the rich can afford to buy. Further, the coastal shrimp industry depletes fishing grounds, salinates drinking water, de-

> stroys mangroves, and displaces fisher people and agriculturists who depend on these resources for their livelihood. In addition, the North American Free Trade Agreement has shifted polluting industries to the coastal belts of the developing nations, threatening the very lives of small fishing communities who are totally dependent on fishing and fishing alone. All over the world the victims of globalization—the small fishing communities—are realizing the need of coming together to establish our right to life. We want to establish our right as persons. The World Forum of Fish Harvesters and Fish Workers is the result of this realization. The forum is involved in a campaign to establish the right of the fishing communities to own bodies of water, including seas and rivers, fishing implements, and distribution of the catch. Management of the distribution of the catch should be done by the fisherwomen.

Kocherry continued on to mention the victories that the networks of fisher people have won against destructive, large-scale industrial fishing not only in India, but in Senegal, in Pakistan, in Brazil, in South Africa, in Canada, and in the United States.[41] In the United States, thanks especially to the wives of fisher people in Gloucester, Massachussets, they succeeded in banning factory trawlers through legislation. "Development," he continues, "is not achieved by conquering and enslaving, not by accumulating and centralizing, not by displacing peoples and destroying cultures. Development consists instead in the capacity to integrate and work together." The convergence with the cause of the populations attacked and expelled by the construction of big dams is evident. Kocherry recalled how already thirty-six hundred megadams had been constructed in India, causing the expulsion from their lands and displacement of fifty million natives, tribal and fisher people, thus proving to be a means of mass destruction rather than development. He argued that these victims want to create a new development paradigm, in which native skills and technologies are enhanced, where small is accepted as beautiful, and sustainability and simplicity become a way of life giving due respect to native cultures.

The concerns of the fishermen communities and those of the populations struggling against the dams justifiably find a common ground in their opposition to the mega-works that proliferate due to this distorted approach to the sea and the rivers. A good example is the story of Umbergaon—once a quiet and obscure fishermen village on the coast of Gujarat in India, today well-known by activists as the home of Colonel Pratap Save, a victim of the struggle for more humane development.[42] Save, who retied in 1995, had decided to spend his old age in that place, working the land of his family. But the village became the object of a mega-plan for the construction of a large industrial port. As soon as the news of the project arrived, Save became one of the first staunch opponents. He formed with others Kinara Bachao Sangharsh Samiti (Save the Coast Action Committee), which promoted various nonviolent actions to block the survey work for the construction of the port. It was feared that it would destroy agricultural activities and the very profitable traditional fishing. Contrary to the government's assurances, people feared that they would become landless and homeless. The actions organized around this issue brought together the populations damaged by the dams, particularly the activists of the Narmada Bachao Andolan, and the coastal population of the area around Umbergaon that lived off fishing and agriculture. Tension grew in the clashes with the government, and the police arrested the activists, among them Save who, taken away in the night, died a few days later. The cause of his death, according to his family, was blows he received on his head. The other activists as well maintained that it was a homicide perpetrated for the benefit of the American company that was to build the port, which after these events pulled out. But people still fear that the project will start again. Undoubtedly, the death of Colonel Save strengthened instead of weakening the determination to oppose the continuation of the work.

Significantly, Father Kocherry himself, at the head of the fishermen movement, decided to go on a hunger strike in Umbergaon to highlight a number of issues. This action aimed at putting pressure on the government so that it would take some initiative against foreign fishing fleets, ban industrial aquaculture, provide fishermen with an adequate quantity of subsidized fuel, and bring to trial those accused of being responsible for the death of Save. The hunger strike came to an end when the government promised that it would move on the question of the fuel, and that it would ban the import of fish into the Indian market. With regard to all the other questions—starting with the communities' right to keep in their hands the resources most crucial for their livelihoods, instead of seeing them privatized and wasted for the profit of the few—the struggle continues. The recomposition of the social subjects involved in the struggle, the type of action, the response of both the companies and the government, describe a situation that is being repeated more and more frequently in India as in other countries. It clearly demonstrates on the one side the growth of a people's front determined to defend an economic and social system that with its resources and culture guaranteed their livelihood, and on the other both the hardening of the companies' responses but at times also their retreats.

Another significant struggle occurred in Mumbai, where on the island of Gorai developers built India's largest amusement park, called EsselWorld, by appropriating public waters and destroying seven hundred acres of mangroves, in violation of the Coastal Regulation Zone.[43] The company built a pier and employed ferries that destroyed the fishing grounds as well as the nets and the boats of the fishermen in the rich bay. On the island live five thousand fishermen who depend on fish for their survival and are opposed to the construction of the park. Opposing this project has meant engaging in several bitter clashes with the police, involving large contingents of police forces mobilized against the hundreds of women fish vendors who have participated in the struggle. But in the end a negotiation began with the participation of the fishworkers and women fish vendors. Here too the strength of the movement, of the men and the women, was felt. Kocherry, who, owing to his commitment, succeeded in becoming a member of the Agency for the Management of the National Regulation Zone, writes in a communiqué related to this mobilization that

> the amusement park must be seen in the context of globalization. Money and profit are the only thing that is valued. To make money the wealthy can expel people and destroy the

> environment. The means of subsistence of the populations are of no concern to the ruling class, the wealthy and the police. The ruling class is selling out what is public for the convenience of the rich. Natural capital is destroyed and the fishermen are expelled.[44]

The community has been steadfast in continuing the struggle to remove the pier and the ferry of EsselWorld, both of which stand to this day.

The tragedy that is unfolding in the world of artisan fishing is very serious, both because of the increasing extinction of fish resources and the destruction of fishing areas due to the proliferation of the industrial fleets and their catching methods, and because of the associated huge loss of jobs among those, men and women, working at different tasks in this sector. Fishing, in fact, provides millions of jobs; they include: the processing of fish; its transport and sale; the construction and upkeep of the boats; and much more. And this is true from Canada to Sri Lanka to Thailand. Moreover, the big trawlers, not only foreign but national as well, often move without observing the limits within which fishing is authorized, or even without licenses. Not only do countries of the Global South generally lack adequate means to monitor the passage of foreign vessels in their waters, they also suffer from widespread corruption. It is the case that many navigate without paying attention to the rules, so that they collide with the smaller boats of artisan fishermen. Consequently, there is a demand for transparency in licensing, for otherwise, as the fishermen point out, it is difficult to proceed toward sustainable development.

The extroversion of development, that is, its strong orientation toward export that is required in the context of the neoliberal globalization promoted by the International Monetary Fund and World Bank, implies that the governments of the South encourage fishing for export (or directly authorize fishing by foreign fleets, or in joint ventures, in their waters) to sell fish to Northern countries in order to acquire hard currency. This, however, is destined to pay for external debt; thus the North receives both the fish and the hard currency. By contrast, still in the name of neoliberalism and structural adjustment, the populations of the countries of the South not only see the subsidies directed to fishing for local consumption eliminated or reduced, similar to what takes place in agriculture, but also see cuts in basic services, such as education, health, and food itself. As is well documented in a report by the association Development and Peace, which is very committed to supporting the organizing done by fishermen in Ecuador, where fish is basically used to acquire hard currency, two hundred thousand Peruvians under five suffer malnutrition, and about half of pregnant women suffer from anemia.[45] Senegal too has adopted a similar fish policy, jeopardizing the availability of this primary source of food to the population. Indeed, according to data issued by the FAO, as we again read in the Development and Peace document cited, fish, mollusks, and crustaceans represent 29 percent of the animal proteins consumed in Asia, 19 percent in Africa, and 8 percent in Latin America. Along the coasts fish is less expensive than chicken and beef. Thus it traditionally constituted the main source of animal proteins and now is disappearing and becoming more expensive because of industrial fishing. Moreover, a third of what is fished, that is, thirty million tons, is not destined for human nutrition but to feed animals, including farmed fish itself. To produce three kilos of salmon farmed with aquaculture it takes 2.7 kilos of fish feed, and to produce these it takes 15 kilos of fish. This is an enormous waste. Seafood has become one of the more globalized economic sectors. About 40 percent of what is fished is traded at the international level at the expense of sales at the local level. Some food multinationals, such as Unilever, Kraft Food, and Nestlé, control a large part of fish stocks.

SELF-ORGANIZATION INITIATIVES

Against this state of affairs, which is increasingly undermining the rights to nutrition and life of coastal communities, not only has a resistance been mounted, aimed at safeguarding the traditional methods of fishing, but a self-organizing effort has been generated, through cooperatives or other collective forms of organization, to determine a development from below, catering above all to the need of those who risk becoming the main victims of industrialization and industrial gigantism. Also in the report cited, to which we refer as a source for the following information, it is mentioned that in Japan fishermen cooperatives self-manage inshore fishing, establishing fishing areas and deciding with regard to occasional conflicts. In the Philippines, an initiative of the community of the Bay of Panguil, on the island of Mindanao, aimed at self-managing an area of 18,500 hectares along 116 kilometers of the coast, has produced remarkable improvements since 1990. Mangroves, an essential vegetation for swampy regions that provides habitat for many wild species, have been replanted; rules for fishing have been established; and artificial coral barriers have been built. The whole thing has constituted a remarkable improvement for the more than 450,000 persons who live in the region. In Nova Scotia, 150 fishermen of the Bay of Fundy have joined to self-manage their fishing. Instead of being assigned individual fishing quotas by the federal government, they have constituted the Fundy Fixed Gear Council to self-manage all their quotas together. The fishermen community decided to work together after recognizing that the resources were limited and a communal approach would be the best solution for good management. In Western Africa, the CNPS (National Collective of Artisanal Fishers of Senegal) has succeeded in participating in the negotiations for a new fishing agreement between Senegal and the European Union. But behind this achievement there is an exemplary story of self-organization. Traditional inshore fishing guaranteed an adequate provisioning of food for tens of thousands of men and women living along the Atlantic coast of this country. It employed sixty thousand fishermen, generating more than two hundred thousand jobs through the activities connected with both the processing of the fish product (salting, smoking) and trading. Many of these jobs were done by women, who invested the income derived from them in the betterment of their families' living conditions and the community's infrastructures(schools, houses, hygienic structures, and other things), making an important contribution to local development. This overall situation was seriously threatened by the advent of neoliberal globalization and with it the big fleets coming from other countries that preyed on the sea. CNPS took off on the basis of a great solidarity between men and women. This collective, created in 1987 and counting as many as 12,500 members from fourteen local fishers' communities, is guided by three great principles: to protect the interests of inshore traditional fishing; to improve living conditions; to promote safety on the sea. It is also consulted by the government, but agreements with foreign partners continue to threaten traditional fishing. Women, therefore, strive to strengthen their fishing-based economies by organizing in groups of savings and credit, enabling them to obtain small sums of money to better develop market programs.[46] The common concerns generated by the arrival of the big trawlers preying on the sea and creating safety problems, because of frequent nightly collisions with small boats, have promoted encounters among fishermen communities, first between Senegalese and those from Canada, and also from Madagascar.

In Canada, the United States, Japan, and Norway, where the crisis has forced fishermen to reduce their crews, women have been reintegrated into the labor process on the boats. Women's participation in the work activities surrounding fishing is generally underestimated, although it is extremely important not only for its size, but also because the income it provides is crucial, enabling men to pay for a crew and go fishing. Here too women active in the fish sector are on the move to promote their interests by building connections and taking advantage of the availability of credit.

Another major example of self-organization by fishermen comes from Agri-Aqua in the Philippines, a coalition that links the land and the sea, farmers and fishermen. Despite the fact that the waters surrounding this country have always been among the richest in fish resources, this wonderful patrimony, a guarantee of livelihood for future generations, is at risk of being destroyed by industrial trawling. Some fish stock have already been depleted, some species have disappeared. Only 5 percent of the coral barrier is still in a perfect state; 77 percent of the mangroves have been destroyed. Agri-Aqua has constituted itself as a coalition of farming and fishing communities of Mindanao, with the objective of safeguarding the living conditions of these communities and building sustainable development from below. The goal is above all to reconstruct the ecosystem. After some of its members studied reforestation techniques it succeeded in restoring the mangrove forests that, after some years, again cover hundreds of hectares, and with the vegetation the fish have come back as have the mollusks, the crustaceans, and the birds, which had hitherto disappeared. A shrimp farm was planned to provide food for the population, and also for trade, so as to generate some monetary income. Some members of the coalition have become part of a management committee of water resources in the region, to promote legislation effectively protecting this patrimony. Other communities are following this example, aiming, above all, at restoring the ecosystem through reforestation with mangroves.

The following are the ultimate objectives of Agri-Aqua, which bring together in solidarity fishermen and farmers: land reform; water reform; sustainable development; equality among the sexes; popular participation at the local government level; and the return of ancestral lands to indigenous people. Around these issues this coalition has grown and represents a great movement for change. It is also a model of practices for fishermen communities in areas of the North, who also depend for their work and livelihoods on the preservation of natural resources. For this reason encounters and connections have been built between Agri-Aqua and fishermen of the Canadian provinces of Nova Scotia and Newfoundland. From the Philippines to Canada, from South to North, a political recomposition is taking place to defend the sources of life, bringing together fishermen, farmers, women, and indigenous people.

NATIONAL FISHWORKERS FORUM AND THE WORLD FORUM OF FISHER PEOPLES

As we can see by considering the moments of struggle and self-organization discussed above, the problematic of the fishermen practicing traditional fishing in India and in the world corresponds very closely to that of traditional farmers: the safeguarding of biodiversity, which is threatened in this case by industrial fishing and farming, and is the basis, instead, of their economy, livelihood, nutrition, and health; the right of access to the sea and the resources that it contains; the right to preserve methods of fishing capable of ensuring the reproduction of the fish patrimony in all its wealth and meeting first of all the needs of the coastal populations. The National Fishworkers Forum (previously National Fishermen Forum) made its voice heard in India especially in the 1990s with strong initiatives, first of all through national strikes that, as we said, involved a great part of the country. However, it has had a consistent history of coordination and support in the fishermen's communities' struggles against industrial fishing and fish farming starting in the 1970s. In this history, a leading role was played by the movement of fishermen of Kerala in southern India. We must remember, in this context, that Kerala represents an exceptional story with regard to governments.[47] With few exceptions, Kerala has always been governed or led by a left party or coalition, promoting and to some degree actually realizing an economic and social project that counters the extreme poverty that has plagued the other states of India and many developing countries. It also eliminated illiteracy, as 100 percent of the population became literate. With the support of many grassroots organizations, the Marxist parties have promoted development sustained by a good welfare system. It was in this context that great ecopolitical battles emerged, which were at the same time battles for economic survival and for the defense of one's culture and identity, with which Marxist parties that had found their bases in the urban industrial unions and the middle class were unprepared to engage. These battles include: the struggle for land, waged by indigenous communities expelled from the forests; the struggle to continue traditional fishing, waged by artisan fishermen; the struggle of workers in nonunionized sectors to protect the sources of their livelihood, such as rivers and other waterways and forests, and again the struggle against industrial pollution and the dumping of urban waste; and the resistance to mega-projects that destroy life-supporting resources and with them the possibility of subsistence. When in the 1960s mechanized trawling began in front of the Kerala coasts, the fishermen of this state, who used traditional methods, immediately verified the damage they suffered through the reduction of what they fished, a consequence of the negative impact of large-scale fishing on the productive capacity of the marine ecosystem, and they began their resistance. In 1979, they organized the

Kerala Independent Fishworkers Federation (KSMTF), probably the largest union in Kerala unaffiliated with any political party. Shortly after its formation, tensions began to emerge in this largely Christian organization between traditional religious elements and politically progressive and modernizing ones. In 1982, the KSMTF split up, but the progressive elements, under the leadership of Father Thomas Kocherry, kept the name KSMTF and almost all the members, who, in time, increased in number. Thanks precisely to the fishermen's struggles, the government banned trawling along the coasts of Kerala during the monsoon season, a period of reproduction from June to September. Later, this movement spread at the national level, becoming, still under the leadership of Father Kocherry, the National Fishworkers Forum (NFF). Its intention was to unify the struggle of the different movements of traditional fishermen working along the Indian coasts into a nationwide network. There were three main issues. First, the struggle against motorboats using giants nets managed through joint ventures with foreign multinationals that, preying on the sea, deprived the fishermen of their means of livelihood. Second, the resistance to large-scale fishing, which destroys the biological diversity of the fishing zones along the coastal areas and the high seas. Third, the possibility of realizing alternatives to large-scale industrial fishing, which is an obstacle to production on a small scale and also ignores the needs of local populations. These issues were combined with the need to have potable water in the village houses, schools, and other basic infrastructures, as well as pensions for all fishermen at the age of sixty. The forum above all promoted the formation of cooperatives exclusively managed by fishermen and village women without any interference from corporations and party organizations. Together with the movements of fishermen from other countries struggling around the same problems and feeling the same needs, the NFF created the World Forum of Fish Harvesters & Fish Workers. At its founding conference, in Delhi in 1997, in which about sixty organizations of artisan fishermen participated, the foundations were set for the formation of a global coordinating committee.

Thomas Kocherry and the Canadian François Poulin, as assistant coordinator, led a first provisional coordinating committee. It was decided, among other things, that November 21, the day of the event, should become World Fisheries Day. The three years that followed this meeting were marked by divergences on how to understand the question of leadership, how to manage a social movement, and about internal democracy. But by the next meeting, held on October 6, 2000, at Loctudy, France, a consensus was reached on the need to build an organization capable of opposing large-scale industrial fishing. After three years of work this important meeting, in a small village in Brittany, marked the true takeoff of this movement at the global level. Participants included twenty-one organizations from sixteen countries and 250 participants, delegates, observers, and listeners representing thirty-two countries.[48] The constituent assembly had to validate two documents to create the basis of a truly global organization capable of decision-making and action: one for the constitution of the forum and a political one devoted to presenting an overview of the demands and strategy common to all the organizations represented. The forum had decided to create egalitarian national delegations (one man, one woman) and to assert the fundamental role of women in the fishing sector, transcending a strictly economic approach to integrate it with a social, familial, and cultural dimension. It gave itself a statute (see Appendix 2) summing up the forum's basic choices. The preamble says the following:

> Gathered in Loctudy (France) under the banner of the World Forum of Fisher Peoples to better defend our work, our access to resources, human rights, the fundamental rights, social justice, the rights, duties, and culture of our communities, affirming that the ocean is a source of life, determined to ensure the sustainability and inexhaustibility of fishing and marine resources for present and future generations, we fishermen coming from every part of the world, today, October 6, 2000, adopt the present Statute and solemnly express our will to respect all its dispositions.[49]

One of the objectives of the statute was to recognize, sustain, and improve the role of women in the economic, political, and cultural lives of the fishermen's communities, ensuring their egalitarian participation in the coordinating committee, correspondence with the egalitarian participation guaranteed to them in the decision-making bodies of the Karnataka State Farmers' Association.[50] The document puts forward many objectives. All the associations and organizations representing the fishermen and their communities belonging to the forum, a possibility denied to transnational industries and large corporations (cf. *article 2* in Appendix 2), must commit themselves to protect the fishermen communities, the resources, and the habitat against any type of threat posed by pollution and industrial aquaculture, and irrational exploitation through destructive fishing methods. Particular attention must be given to coastal communities, to their preservation and valorization. Food security must be guaranteed on a world scale by ensuring the renewal of stocks, and consequently fish, with their beneficial properties to human nutrition.

In order to realize these objectives, the forum gave itself an organizational structure composed of: a general assembly formed by all the delegates of the participating organizations; a coordinating committee composed of two coordinators, a general secretary, a treasurer, and continental representatives, with the tasks of planning and managing the activities of the forum, realizing its objectives, and promoting continental forums; and five continental forums constituted of active members, with their own autonomous management.

At Loctudy the participants also decided to adopt the principles of People's Global Action, a transnational network that coordinated broad social struggles during the cycle of protests of the antiglobalization movement, and had worked collaboratively on plans for alternatives to capitalism at the local level, implementing models for organizing that favor decentralization and autonomy, that meet the real needs of local communities, and that are sustainable for the sea and those who live there. These principles are: a complete rejection of feudalism, capitalism, and imperialism; a complete rejection of trade agreements and of institutions and governments promoting a destructive globalization; a complete refusal of all forms and systems of domination and discrimination, among them patriarchy, racism, and religious fundamentalism in all creeds, so that the dignity of all human beings is recognized; a politics of dialogue, so that capitalism is not the only factor guiding the economy; an appeal to direct action and civil disobedience in support of the struggles and movements mobilizing for the right to life and the rights of oppressed peoples and to build local alternatives to capitalism; and the realization of models promoting decentralization and autonomy.[51] These are all principles that resonate with the statutes put forth by the NFF that, in any case, is primarily oriented to the protection of the sea and its resources.

Another important objective defined in the works of the founding assembly is to commit the continental forums, on World Fisheries Day, November 21, to the organization of demonstrations and strikes aiming to sensitize the public to the problems cited by the World Forum of Fisher Peoples. In particular, the aim is to stress the rights of local communities to have access to resources and to be able to manage them autonomously. For the organization of November 21, a communiqué of the general secretary states:

> The World Fisheries Day is a way of expressing our identity.Every country must express a political, economic, and social commitment in favor of a campaign of consciousness-raising. . . . The fishing communities depending on these activities for their livelihood should own their resources, seas, lakes, ponds. They should manage what is fished [and] its distribution, and receive subsidies. . . . Everywhere there is competition for the waters. . . . We need to organize in collectives and cooperatives.[52]

The communiqué foresees the drafting of fliers and posters informing civil society about the struggles waged by members of the forum, coordinated on a world scale by the committee. Indeed, in November 2001, a global strike was organized involving the whole fishing world to oppose the looting of the seas. Fishermen put down their tools to say "enough" to the floating fishing industries, to say "stop" to the pollution of the seas and to industrial monoculture. It was the largest strike affecting the fishing industry in the history of humanity.[53] The targets of the forum were the pelagic trawling nets and the fishing methods that most compromise the health of the sea. On this point the associations agree with Article 6.1 of the FAO's Code of Conduct for Responsible Fisheries (1995). The article recommends the use of methods of fishing that ensure the effective conservation of the ecosystem, in particular to prevent overfishing. The forum, however, does not approve of the part of Article 6.1 that recommends that trade in fish and fishery products should be conducted in accordance with the principles, rights, and duties established by the World Trade Organization (WTO).[54]

At the second World Social Forum, held in Porto Alegre in January 2002, the positions taken in the statute and on World Fisheries Day were reconfirmed. In agreement with the other five thousand movements and organizers coming from five continents, it was decided to continue the struggle against the dominant models and the privatizations carried out by the multinationals that with 10 percent of the population take 80 percent of the world's wealth, leaving the remaining 90 percent of the world's population to share the 20 percent that remains after the feasts of the former.[55]

Following the development of the World Forum of Fisher Peoples, we find that proposals have not been lacking, even at the World Summit on Sustainable Development held in Johannesburg in August 20002. Some of the statements affirm the important right to live and survive on this planet with honor and dignity. This is considered possible only if there is social and economic justice, truly sustainable development, and a harmonious relationship with the environment guaranteeing biodiversity to future generations. These are concepts that are more and more frequently reaffirmed in summits, forum, and conventions. The forum has requested the implementation of Chapter 17—Protecting and Managing Seas, Oceans, Their Resources and Coastal Zones—of Agenda 21, which was drafted in Rio in 1992. However, the action plan for the international community had not yet contemplated juridical obligations, and the text remained a purely programmatic gesture.

At Loctudy, disagreements concerning the forms of organization, how social movements should act, the relationship with power, as well as fishing methods led to a split, and then to a constitution of two forums. The Asian delegates and the majority of the African delegates followed Kocherry, giving life to a second World Forum of Fisher Peoples.

Along with many other popular movements, representatives from the fishermen movement, either as the National Indian Forum or as the World Forum, did not participate in the World Social Forum held from January 16 to January 21, 2004 in Mumbai. Several social groups judged the role of the Indian NGOs in the management of the social forum negatively.[56] The absence of a genuine prospect of struggle against global capitalism led to the exclusion from the forum of many important grassroots movements and the self-exclusion of others. Among the movements excluded or self-excluded—in addition to the fishermen's movements already mentioned and Maoist organizations, both of which are important in India and in Asia—were the Sem Terra of Brazil; the National Federation of Farmers Organization, undoubtedly the most powerful opponent, singularly considered, against the WTO; other important farmers' movements from Asia, such as those from the Philippines and Nepal; a large part of the National Alliance of Peoples' Movements (NAPM); and the National Association of Landless and Agricultural Workers. These movements, all radically anticapitalist, organized other moments of encounter and demonstration outside the social forum, such as the blocking of the railroad station, as a form of protest against the invasion by industrial trawlers. The Mumbai Resistance 2004 and the People's Movements Encounters II, for instance, were constituted outside the social forum.

The year 2004 marked another important step for the fishermen's movement: the decision by the National Fishworkers' Forum of India and the World Forum of Fisher Peoples to participate in the elaboration by the International Labor Organization (ILO) of a regulation for the informal labor sector of the fishing industry. It was probably the first time that the ILO had to be concerned with such a question. Informal work produces, as we recalled earlier, 25 percent to 39 percent of the total catch in India, but 80 percent to 90 percent of the ten million fish workers in this country depend on that percentage. If marine resources are not preserved, there will be no security for these workers. This is why the WFFP decided to participate in the work of the ILO, to reaffirm, first of all, the crucial organic connection between these resources and traditional fishermen, between the preservation of the marine ecosystem and the possibility of work and livelihoods for the coastal communities. Accepting the need, in order to be able to participate in the discussion, of adopting the viewpoint of the work, regardless of the type of vessel considered, rather than the viewpoint of the caste or the community, the participants debated every issue, starting with whether they should draft conventions or recommendations or both. Through a questionnaire circulated by the ILO they analyzed the basic issues, such as: the necessity to have written contracts and lists of the people onboard fishing vessels; the need to have identity cards and stipulated rights with regard to repatriation and recruitment; the need for laws and regulations concerning accommodations onboard, food, and potable water; the need for norms ensuring medical treatment, including first-aid treatment and the provision of kits (such as life jackets) in good condition for emergency situations; and the need for legislation ensuring welfare provisions already provided to other workers, above all pensions and insurance, all the more necessary considering the high risks involved in this type of occupation.[57] They focused their attention, above all, on: the need to establish a minimum age to work onboard; a minimum number of hours of rest proportional to the hours of work; the need to ensure the education of boys, through the introduction of flexible school time, since delaying working on boats by years makes them more vulnerable to seasickness and the risk of not acquiring a sufficient confidence with the marine environment. They also asked for a minimum wage for those receiving pay, with special attention to the work done by migrants and tribal people.[58] In another document, this time written by the WFFP, *Toward a Fisheries Policy in India*, the organization calls for the development and implementation of a fisheries policy in the country, taking into account the basic needs of the coastal populations and their organic relations to the fishery resources.[59]

This network insists, above all, on the need and the will to continue the struggle against the industrialization of fishing, to defend the right of traditional fishermen to own in common the nation's fishery resources, as they depend on this activity for their livelihood. In conformity with what has been mentioned in other documents, here too it is restated that 90 percent of the fishermen catch about 30 percent of the three million tons that are fished. But these fishermen are those who carry out labor-intensive fishing, done with traditional methods, and they depend on this activity and therefore depend on the safeguarding of the fish patrimony for their survival. A legislative change—they argue—will have to first block the globalization process that guarantees the free movement of capital. "We struggle," the document states, "for the free movement of workers across the world." It is regrettable that more than sixty years after independence, India is still waiting for a fishing policy regulating its seven thousand kilometers of coasts and the correspondent exclusive economic zone (EEZ).[60] After the great national strikes of fishermen, the Murari Committee was constituted by the central government in 1995, composed of sixteen parliamentarians and all the ministers organically linked to the sector, both for marine and internal waters, and also by six representatives of various stakeholders. With the formation of this committee the fishing question was dealt with for the first time. The twenty-one recommendations that emerged from its work were important, because they touched on all aspects of the problem, and they were also accepted by the government on September 27, 1997. Had they been implemented, a fishing policy for India would have been realized. But this did not happen; as a result, up to this day nothing has been done on this matter. The National Fishworkers Forum made another attempt by launching a mobilization under the slogan "act or perish," centering on forty-two demands and the restated need for a regulation safeguarding the marine patrimony. For a sea in which catamarans and large industrial vessels fish, the forum demanded the implementation of the Marine Fishing Regulation Act of 1978 that was aimed at protecting three basic aspects of fishing: the economic life of traditional fishermen; the conservation of fish resources; and the enforcement of law and order on the sea. But, the document maintains, politicians and bureaucrats have prevented the implementation of this act to promote their interests, and this has opened the door widely to the unrestricted arrival of large trawling vessels that destroy the seabed. Similarly, the introduction of high-sea fishing in 1991, together with other policies typical of neoliberal politics, was a move in this direction, in the absence of a well-defined fishing policy. The document denounces the thousands of large fishing fleets in the world that have exhausted the reserves of all the oceans, except for those of the Indian Ocean and the African seas. These vessels are totally or partially useless. The European Union and the United States give them large subsidies to make them go outside their waters, to make them go to the Asian or African waters. The Indian government wanted to import twenty-six hundred of these vessels, but it clashed with the opposition of the fishermen, who succeeded in blocking this initiative. Now, this same government wants to change the recommendations of the Murari Committee and import these vessels at any cost, which is why, the documents reports, it is important to keep the opposition alive. From this perspective, in which the conflict between life and profit is becoming more and more dramatic, it is significant that the growth of the mobilization also leads to an increasing recognition of the crucial role of women in this sector, as does the decision to end any form of discrimination against them. All this is also restated in the statement of the National Fishworkers Forum demands, together with the need to ensure that women have, in this context, adequate representation, both in the cooperatives and on the different committees and commissions. It is emphasized that freedom of movement for workers would also resolve the problem of overcrowding in some fishing situations. The need to put an end to all forms of pollution of the sea, including nuclear pollution, is reaffirmed. There is a call for welfare policies ranging from subsidies for fuel, promised many times but never delivered, that would enable fishermen to venture into deeper waters, to pensions, to some form of insurance and credit that would guarantee more security. The document denounces the expulsions of fishermen communities from the lands where they have always lived, which are justified with specious excuses and, in the same territories, open the door to timber companies or other businesses. With the excuse of protecting the forests or the sharks or the turtles or some marine sites, tens of thousands of fishermen are constantly expelled. The Wildlife Protection Act has been used to "protect" fish against human beings, whereas this law concerns the protection of animals in the forest, not fish, as these fall under the jurisdiction of the Marine Fishing Regulation Act, which aims at protecting the fish patrimony together with the fishermen's communities. It is restated that the government must commit itself to preserving marine resources, while there is opposition against the government's effort to intensify the productive capacity of fishing, particularly by acquiring, as we have mentioned earlier, large industrial vessels. Denouncing the fact that the fishermen's communities continue to live in inhuman conditions, above all because of the lack of space for houses, the document asks all the states affected by fishing activities to commit themselves to ensuring the availability of food, housing, education, health care, potable water, and the basic infrastructure that a village needs.

It also asked, in conformity with one of the twenty-one recommendations of the Murari Committee, that a democratically constituted and democratically managed agency be formed to supervise the implementation of the Marine Fishing Regulation Act. It asserts that the government not only has not abided by the verdict of the Supreme Court that ordered the removal of all industrial aquaculture farms along the coast of India, subject to the regulation, but it has, instead, introduced the Aquaculture Authority Bill to legitimate the industrial farming of shrimp in these areas. Worse yet, this question would be placed under the responsibility of the minister of agriculture.

Yet the movement of fishermen has undoubtedly achieved very important objectives. Through the network it has built it is forcefully promoting, at the international level, the institution of a labor regulation for fishing, even an informal one. In the meantime it is exerting great pressure for effective legislation in order to protect the marine ecosystem and the immensely abundant fishery patrimony it represents. This network experiments with self-managed forms of development from below, with creative and not destructive development, respectful first of all of the needs of the coastal populations and valorizing the resources of their habitat. Preserving the spontaneous reproduction of life in the seas, lakes, and rivers is in fact the greatest guarantee of security for all. The flyer distributed for World Fisheries Day in 2004 states:

> On November 21st, we are preparing once again to celebrate World Fisheries Day, a day dedicated to the fishermen of the world so that they realize that they are at the center of the development and management of marine resources . . . we are experiencing a crisis because resource fish are overutilized by factory-like ships, by the scraping of the seabeds with seine nets, by excessive fishing and pollution. We must continue the struggle against the oversizing of boats, against excessive fishing, against destructive equipment and pollution of all types. The waters are polluted by domestic and industrial trash, by plastic materials and nuclear waste. If this continues, all the waters will become extremely polluted, and life on the planet will be in danger. We cannot destroy life on the planet in the name of development. The lives of millions of people depend on these waters. It is our responsibility and our duty to safeguard them for the good of humanity and the planet. Let us go forward together saying: "Let's protect the waters, let's protect life. . . . We will take care of you, our mother ocean, and the marine resources, until [for as long as] we have a breath of life."[61]

NOTES

1. United Nations World Summit on Sustainable Development, *Fighting Poverty*, August 26–September 4, 2002, Johannesburg, South Africa, available at: www.wws.it/ambiente/sostenibilità/pianoazione_1.asp.
2. F. Carlini, "Ipocritipescatoriinacquealtrui," *IlManifesto*, February3, 2002, in www.ilmanifesto.it/php3/ricview.php3?page=/terraterra/archivio/2002/febbraio/3c5e75coce563.htmI&word=pescatori.
3. "Accord de pêche Ue-Mauritanie, " *Le Courrier ACP-UE*, n. 191, 2002.
4. *Rischiano il collasso gli stock ittici dell'Africa Occidentale*, June 26, 2002, available at: www.wwf.it/news2862002_4229.asp.
5. "*Ecologist Asia*," 3 (July–August 1995); V. Shiva, *Stolen Harvest: The Hijacking of the Global Food Supply* (Cambridge, MA: South End Press, 2000), 39–40.
6. Notoriously, the green revolution has led to a growth in agricultural productivity of the major crops, such as wheat, rice, and corn, through the utilization of big mechanical means, chemical inputs, and the adoption of hybrid varieties. The agronomist Norman Borlaug, who elaborated a particular hybrid variety of wheat and cultivated it starting in the 1950s, of the twentieth century is considered the father of the green revolution.
7. V. Shiva, *Monocultures of the Mind: Perspectives on Biodiversity and Biotechnology* (London: Zed Books, 1993).
8. G. Giovannelli, "Ogni gamberetto mille poveri in più," *Vita*, May 26, 2000.
9. An industrial shrimp farm requires 120,000 cubic meters of seawater per hectare. This water must then be diluted with fresh water drawn from subterranean aquifers, a process which makes the latter vulnerable to the input of salted water. (V. Shiva, *Stolen Harvest*)
10. On the attack against the commons, above all the land, starting in the 1980s in the context of neoliberal politics, that has been interpreted as a cycle of "new enclosures," see the thorough analysis that the Midnight Notes Collective has produced, triggering a vast international debate. Among the various publications by the Midnight Notes Collective, see *New Enclosures* (Brooklyn, NY: Autonomedia, 1990). Concerning the contradiction between capitalist development and the commercialization/destruction of nature, defined and analyzed by James O'Connor as the "second contradiction of capitalism," a leading role has been played by the journal *Capitalism Nature Socialism*, under the direction of this scholar. The Italian version of this journal has been *CNS Capitalismo Natura Socialismo*, later called *CNS Ecologia politica*. As for the present debate on the commons, we refer to the online journal *The Commoner*, which has

produced over a dozen issues and which has hosted a broad and interesting discussion in May 2001.

11. M. Shanahan, *Appetite for Destruction*, March 22, 2003, available at: www.theecologist.org/archive_article.html?article=376&category =88.

12. V. Shiva, *Stolen Harvest*.

13. M. Shanahan, *Appetite for Destruction*.

14. P. Giansanti, "Afghani, otto anni. Valore: 15 chili di gamberetti," *Sette*, Supplement to the *Corriere della Sera*, November 29, 2002. On child labor in slavelike conditions, see K. Bales, *Disposable People: New Slavery in the Global Economy* (Berkeley: University of California Press, 1999).

15. V. Shiva, *Stolen Harvest*, 43.

16. Ibid., 43.

17. Ibid., 49–51.

18. V. Shiva, *Biopiracy: The Plunder of Nature and Knowledge* (Cambridge, MA: South End Press, 1999)

19. M. Shanahan, *Appetite for Destruction*.

20. www.earthsummitwatch.org/shrimp/national_reports/crhondt.html.

21. F. Gioanetto, "Terra Terra, " *Il Manifesto* June 26, 1999, in www.ilmanifesto.it/php3/ricview.php3?page=/terraterra/archivio/1999/giugno/3b28921564280.html&word=gamberi.

22. M. Shanahan, *Appetite for Destruction*.

23. *Bangladesh: violenze sui contadini*, available at: www. manitese.it/boycott/boyc36/Bangladesh.htm. M. Shanahan, *Appetite for Destruction*.

24. *National Report from Malaysia*, June 1997,in www.earthsummitwatch.org/shrimp/national_reports/crmai1.html.

25. *National Report from Thailand*, January 1998, available at: www.earthsummitwatch.org/shrimp/nationa1_reports/crthai.html.

26. M. Forti, "Terra, Terra," *Il Manifesto*, November 18, 1999, available at: www.ilmanifesto.it/php3/ricview.php3?page=/terraterra/archivio/1999/Novembre/3b2892b45c580.html&word =gamberi.

27. M. Forti, "Terra, Terra," *Il Manifesto*, January 7, 2000, in www.ilmanifesto.it/php3/ricview.php3?page=/terraterra/archivio/2000/Gennaio/3b2892f4caa30.html&word=gamberi.

28. *India: ora I gamberetti non uccidono* piú, available at: www.manitese.it/boycott/borc27/gamberi.htm.

29. M. Forti, "Terra, Terra," *Il Manifesto*, July 25, 1999, in www.ilmanifesto.it/php3/ricview.php3?page=/terraterra/archivio/1999/Luglio/3b289223900c80.htm&word=pescatori.

30. *India: ora I gamberetti non uccidono piú*, available at: www.manitese.

it/boycott/borc27/gamberi.htm.

31. Jagannathan, a member of a high caste, left college to join the struggle for independence, burning his elegant English shirt. On the road to Gandhi he met Krishnammal, an untouchable. Born in 1926 in the village of Ayyankottai, from a Harijan family, the fifth of twelve children, Krishnammal ran away from home when only twelve and was adopted by a famous Gandhian follower, the female doctor Soundaram Ramachandran, who made her study. The two married in 1950 without the approval of their families, and since then they have dedicated themselves to the defense of the weakest, consistent in their thoughts, actions, and lifestyles. (L. Coppo, *Terra gamberi contadini ed eroi* [Bologna: Emi, 2002])

32. "Insieme agli affamati nel segno del Mahatma," *Vita*, May 26, 2000, available at: www.cesvi.org/scuola/pdf/26_05.pdf.

33. *About 9th September Rally*, September 10, 2002, available at: www.laft.org/moduled.php?op=molbad&name=News&file=article&sid=5. Concerning the damages caused by cyclones and hurricanes in India and the coasts of Asia, see V. Shiva, *Water Wars, Privatization, Pollution and Profit* (Cambridge, MA: South End Press), 46–47.

34. These subsidies include: up to 25 percent of the capital invested, or 30,000 rupees per hectare; up to a maximum of 150,000 rupees for new aquaculture farms; up to 25 percent of the capital invested, or a maximum of 500,000 rupees for the construction of shrimp farms of medium-size dimension, with an annual capacity of 30 million or more of seeds per year; 25 percent of the capital invested in food or eggs, or a maximun of 3,450 rupees per hectare for food and eggs respectively; 25 percent of the expenditure for the constitution of a reproduction bank, or a maximum of 150,000 rupees. Lastly, the shrimp farmers benefit from a custom reduction on the import of feed for the shrimp. (V. Shiva, *Stolen Harvest*, p. 55.

35. G. M.George, "Prawn Farms Destroying the Life of Fisherworkers," *Counter Currents,* May 5, 2004 available at: www.countercurrents.org/en-goldy-50504.htm.

36. See the graphic related to "World Production of Fish Catch in 2000" in Chapter two, p. 30.

37. Speech given at the Nineteenth United Nations General Assembly Special Session (UNGASS) for the Revision and Evaluation of the Implementation of Agenda 21, New York, June 23–27, 1997, available at: www.scarpin.com/spn_leg&nor_doc.3.htm.

38. *Inter Continental Caravan 99*, available at: www.rfb,it/icc99/default.htm.

39. The World Trade Organization and its neoliberal politics are notoriously at the center of the contestation of the movement of movements.

40. *Presentazione dei vari movimenti indiani che partecipano alla ICC*, available at: www.rfb.it/icc99/presentazioni_movimenti.htm.

41. Speech by Sophie Prize Winner Thomas Kocherry, June 15, 1999, available at: http://www.sofieprisen.no/Articles/52.html.

42. Rajni Bakshi, "Civil Disobedience in Umbergaon," *The Hindu*, July 22, 2001, in www.hindu.com/thehindu/2001/07/22/stories/13220610.htm.

43. See "India, Fisherfolk Struggle, Mumbai: Background, Action and Arrests," archived on *A-Infos*, July 11, 2000, available at: www.ainfos.ca/oo/jul/ainfosoo162.htm.

44. Ibid.

45. *Development and Peace and the Fisheries*, 1998 available at: www.devp.org/testA/issues/fisheries/htm.

46. For an analysis of the different forms of microcredit and the problematic relative to women's self-organizing in Africa in times of peace and war around this resource see F. Zamperetti and G. F. Dalla Costa eds., *Sharing Trust: Women and Microcredit in Eritrea* (Trenton, NJ: The Red Sea Press, 2008).

47. An essential article on this matter is that by G. Madhusoodanan, "*Il modello Kerala alla prova* d*ell'ambientalismo*," our source for the following observations concerning this state. In *CNS Ecologia politica* 55–56 n. 3– 4 (August–December 2003).

48. The organizations belonging to the World Forum of Fisher Peoples are as follows:

All Pakistan Fisherfolks, Pakistan

Artisanal Fishers Association, South Africa

Association Familiale Maritime, Martinique

Association des Pêcheurs et Fisciculteurs du Mali, Republique de Mali

Bangladesh Jele Federation, Bangladesh

Bigkis Lakas Philipinas Inc., Philippines

Collectif Nationale des Pêcheurs Artisanaux, Senegal

Collectif des Organizations Maritimes Malgaches, Madagascar

Confradias Firmantes Carta De Cedeira, Spain

Fishing Community Network, Pakistan

Katosi Women Fishing Group, Uganda

National Fish Workers Forum, India

National Fisheries Solidarity, Sri Lanka

National Union of Fishermen, Sri Lanka

Penang Inshore Fishermen Welfare Association, Malaysia

Rongomaiwsine Trust Inc., Aotearoa, Pacific

Uganda Fisheries and Fish Conservation Association (Uffca), Uganda

Union Nationale des Pêcheurs Artisans De Guinee, Guinea

Union des Professionnels de la Pêche Artisanal, Benin

United Federation of Labor, Sri Lanka

Source www.wffp.org/index.asp?filer=web3.htm.

49. www.wffp.org/index.asp?file1=web8.htm. Translated from the French by Monica Chilese.

50 *Du 2 au 6 Octobre se tenait a Loctudy l'Assemblée Constituante du Forum Mondiale des pêcheurs artisans qui a aboutit le dernier jour à la constitution de deux forums distincts*, available at : www.alliance21.org/fr/colleges/fish/docs/fishforum_fr.rtf.

51. *Principes de l'Amp*, available at: www.nadir.org/nadir/initiativ/agp/fr/PGAInfos/hamfr.htm.

52. *World Fishery Day*, November 21 2000, available at: www.wffp.org/index.asp?file=web5.htm.

53. *Global Fisheries Strike*, November 21, 2001, in www.wffp.org/index.asp?filer=globalstrikeposter.htm.

54. *First General Body Report*, October 6, 2000, in: www.wwfp.org/index.asp?filer=web3.htm.

55. *Second World Social Forum*, available at: www.wffp.org/index.asp?filer=secondworldsocial forum.htm.

56. O. De Marcellus, "Divisions and Missed Opportunities," *Metamute*, n. 28 (Summer–Autumn 2004).

57. *WFFP and ILO Convention on Labour Standard in the Fishing Sector*. May 9, 2004, available at: www.wffp.org/indexcontent.asp?filer=may904.html.

58. *International Collective in Support of Fishworkers South Indian Federation of Fishermen Societies*, April 2–3, 2004, available at: www.wffp.org/indexcontent.asp?filer=ilo.html.

59. *Toward a Fishery Policy in India*, available at: www.wffp.org/indexcontent.asp?filer=mayo804.html.

60. The exclusive economic zone (EEZ), let us briefly recall, was introduced at the Third Conference of the United Nations Conference on the Law of the Sea (UNCLOS III), which concluded with the ratification of the Convention on the Law of the Sea. It stretches not beyond the two-hundred-mile limit from a nation's coast, including in this territory all the marine environment. For more detailed information, see Chapter 2, note 5.

61. *Journeé Mondiale de la pêche 2004*, available at: www.wddp.org/index content.asp?filei=worldfishfrench.htm (see Appendix 3).

APPENDIX ONE
THE SITUATION OF FISHING IN THE EUROPEAN UNION[1]

The European Union is the third-largest power in the world in the fishing sector. It represents, moreover, the largest market for fishing and aquaculture products. Its nearly one hundred thousand fishing boats catch annually about six to seven million tons of fish, worth about €7.3 billion. We report here the most significant data from the sector by nation.

AUSTRIA
Number of boats: no fishing boats
Overall tonnage: figures not available
Number of fishermen: 150 part time
Employment in the entire sector: 400 full-time jobs, 650 part-time, and about 1,500 seasonal jobs
Annual fishing: 454 tons in Austrian territory
Value of the fishing: €12.7 million for aquaculture and €2.7 million for the fishing on Austrian territory
Average annual consumption of fish per person: 10.6 kilos
Export of fish: 5,451 tons
Subsidies: €15.62 million (1994–1999)
Politics pursued: It fully supports reforms to the fishing sector. It places environmental questions at the top of its policy agenda.

BELGIUM
Number of boats: 142 (148 in1998)
Overall tonnage: 23,085 gross tons and 64.896 kw (1998)
Number of fishermen: 745, of whom 87 are part time
Employment in the entire sector: 1,992 full-time and 151 part-time jobs
Annual fishing: about 30.325 tons
Value of the fishing: €103.4 million (1998)
Average annual consumption of fish per person: 19.6 kilos
Export of fish: 94,272 tons
Subsidies: €59.98 million (1994–1999)
Politics pursued: Devoted to reforms. It wants a sustainable fishing sector.

[1] www.wwf.it/pesca/europa.asp

DENMARK

Number of boats: 4,153

Overall tonnage: 97,664 gross tons (1994–1999)

Number of fishermen: about 6,000 (2000)

Employment in the entire sector: 11,333 (1998)

Annual fishing: about 1.9 million tons (2000)

Value of the fishing: €527 million (2000)

Average annual consumption of fish per person: 23.6 kilos

Export of fish: 1,862,795 tons (2000)

Subsidies: €305.32 million (1994–1999)

Politics pursued: Favorable to a politics that could make fishing more sustainable, using a precautionary approach based on concern for the ecosystem. It would like to have more measures to reduce the fleet and reduce subsidies. It also calls for agreements regulating access to fishing that take into account the needs of local populations and that create policies more coherent with the development of the EU.

FINLAND

Number of boats: 3,762 (2000)

Overall tonnage: 24,170 gross tons (1998)

Number of fishermen: 3,733 (2000)

Employment in the entire sector: 5,563 (1999)

Annual fishing: about 112,000 tons (1998)

Value of the fishing: €28 million (1998)

Average annual consumption of fish per person: 32.8 kilos

Export of fish: 22,204 tons

Subsidies: €60.73 million (1994–1999)

Politics pursued: The Finnish government is favorable to reforms and supports the guidelines that come from the Green Book of the EU.

FRANCE

Number of boats: 8,836 (1998)

Overall tonnage: 209,406 gross tons (1998)

Number of fishermen: 21,664 (1998)

Employment in the entire sector: 53,657 (1998)

Annual fishing: 577,206 tons, 27,008 tons overseas.

Value of the fishing: €932.4 million (1998)

Average annual consumption of fish per person: 24.4 kilos

Export of fish: 447.66 tons

Subsidies: €157,766 (1997)

Politics pursued: It is generally reluctant to accept the changes suggested by the politics of fishing, such as subsidizing or reducing quotas. It prefers to accept regulation rather than reducing capacity.

GERMANY
Number of boats: 2,313 (1999)
Overall tonnage: 69,800 gross tons (1999)
Number of fishermen: 4,370 (2000)
Employment in the entire sector: 46,410 (1999)
Annual fishing: about 254,000 tons (1999)
Value of the fishing: about €190 million (1999)

Average annual consumption of fish per person: 12.4 kilos
Export of fish: 705,000 tons (1999)
Subsidies: €289 million (2000–2006)
Politics pursued: It is very favorable to reforms and would like to have environmental questions placed at the center of the European politics of fishing. It considers the management of the ecosystem a priority, as well as a more balanced restructuring of fleets and resources.

GREECE
Number of boats: 20,243 (1998)
Overall tonnage: 11,933 gross tons (1998)
Number of fishermen: 43,952 (1997)
Employment in the entire sector: 49,525 (1997)
Annual fishing: about 124,386 tons (1997)
Value of the fishing: €458.2 million
Average annual consumption of fish per person: 25.4 kilos
Export of fish: 73,210 tons
Subsidies: €174.98 million (1994–1999)
Politics pursued: It is favorable to defining limits and protected areas. It is uncertain with regard to the activity of other Mediterranean countries and would like to have better collaboration and regulation of Mediterranean fishing.

HOLLAND
Number of boats: 1,040 (1998)
Overall tonnage: 174,344 gross tons (1998)
Number of fishermen: 2,431 (2000)
Employment in the entire sector: 9,557 (1997)
Annual fishing: 546,477 tons (1998)
Value of the fishing: €358.1 million
Average annual consumption of fish per person: 15.4 kilos
Export of fish: 855,736 tons
Subsidies: €33.63 million (1996–1999)
Politics pursued: It is favorable to reforms; it would like sectoral-wide cooperation allowing for the participation of fishermen. It would prefer to face more regulation rather than the reduction of capacity. Generally, it agrees with the reduction of subsidies for fishing.

IRELAND
Number of boats: 1,246 (1998)
Overall tonnage: 61,082 gross tons (1998)
Number of fishermen: 6,424 (1998)
Employment in the entire sector: about 11,300
Annual fishing: 324,843 tons (1998)
Value of the fishing: not reported
Average annual consumption of fish per person: 16.8 kilos
Export of fish: 273,792 tons
Subsidies: €71.21 million (1994–1999)
Politics pursued: It does not want a reduction of its fleet but is favorable to more restrictive technical measures. For example: it advocates the use of nets with larger meshes and the creation of zones where fishing is banned.

ITALY
Number of boats: 16,522 (2002)
Overall tonnage: about 200,000 gross tons
Number of fishermen: 53,000
Employment in the entire sector: 70,000
Annual fishing: 463,400 tons
Value of the fishing: €2,023 million
Average annual consumption of fish per person: 22 kilos
Export of fish: 116,000 tons
Subsidies: €325.50 million (1994–1999)
Politics pursued: It opposes reforms and especially the reduction of fleets.

LUXEMBOURG
Data not available

PORTUGAL
Number of boats: 11,579 (1998)
Overall tonnage: 123,923 gross tons (1998)
Number of fishermen: 27,197
Employment in the entire sector: 40,770 (1998)
Annual fishing: 189,529 tons (1998)
Value of the fishing: €252.4 million (1998)
Average annual consumption of fish per person: 59.8 kilos
Export of fish: 99,011 tons
Subsidies: € 218.17 million (1994–1999)
Politics pursued: It does not want major changes nor cuts to quotas.

Spain

Number of boats: 16,646 (2000)
Overall tonnage: 525,134 gross tons (2000)
Number of fishermen: more than 66,800 (2000)
Employment in the entire sector: more than 470,000 (2000)
Annual fishing: 1,079 million tons (1999)
Value of the fishing: €1,952 million (1998)
Average annual consumption of fish per person: 30.3 kilos
Export of fish: 762,229 tons
Subsidies: more than €2,352 million (2000–2006)
Politics pursued: Favorable to measures reducing bycatch but opposes any reduction of subsidies and capacities of its fleets.

Sweden

Number of boats: 1,968 (1998)
Overall tonnage: 47,000 gross tons (1999)
Number of fishermen: 2,880 (1998)
Employment in the entire sector: 4,932 (1999)
Annual fishing: 330,350 tons (1999)
Value of the fishing: €108.1 million (1999)
Average annual consumption of fish per person: 17.4 kilos
Export of fish: 248,800 tons (1999)
Subsidies: €119.1 million (1994–1999)
Politics pursued: It is favorable to reforms. It would like to reprogram subsidies and have better agreements about fishing permits with emergent countries.

United Kingdom

Number of boats: 8,658 (1998)
Overall tonnage: 253,409 gross tons (1998)
Number of fishermen: about 15,000 (2000)
Employment in the entire sector: 38,454 (1997)
Annual fishing: 400,000 tons arrive in the ports of the country (1999)
Value of the fishing: €318 million (1998)
Average annual consumption of fish per person: 20.1 kilos
Export of fish: 695,649 tons
Subsidies: €244.67 million (1994–1999)
Politics pursued: It is favorable to reforms. It would like to have long-term recovery programs and more regional management, increasing the participation of those working in the sector.

APPENDIX TWO
STATUTE OF THE WORLD FORUM OF FISHER PEOPLES (WFFP)[1]

AIMS

Protect, defend, and strengthen the communities that depend on fishing for their livelihood. Represent the small-scale, artisan fisher's interest at the international level and play the role of global political organ of the same communities.

PREAMBLE

Gathered in Loctudy, France, under the banner of the World Forum of Fisher Peoples to better defend our work, our access to resources, our human rights, our fundamental rights, social justice, and the rights, duties, and cultures of our communities, affirming that the ocean is a source of life, determined to ensure the sustainability and inexhaustibility of fishing and marine resources for present and future generations, we fishermen coming from every part of the world, today, October 6, 2000, adopt the present Statute and solemnly express our will to respect all its dispositions.

OBJECTIVES

Article 1: Objectives

The objectives of the World Forum of Fisher Peoples (WFFP) are:

a) To protect, defend, and strengthen the communities that depend on fishing for their livelihood.

b) To assist member organizations to secure and improve upon the economic viability and quality of life of Fisher Peoples and their communities.

c) To recognize, support, and enhance the role of women in the social, economic, political, and cultural life of the fishing community.

d) To create an understanding of the ocean's resources as a com-

[1]Sources http://worldforumoffisherpeoples3.blogspot.com and www.wffp.org/index.asp?filer=web8.htm. Translated from French by Monica Chilese.

mon heritage of humanity and, through sustainable fishing practices, conservation, and the regeneration of the marine and inland resources and ecosystems, to ensure that it is passed on to future generations.

e) To protect fishing communities, fish resources, and fish habitats such as coastal zones, watersheds, and mangroves from land-based, sea-based, and air-based threats. These include displacement by tourism, pollution (including the use of the sea as a dumping ground for toxic waste), destructive industrial aquaculture, overfishing, and destructive fishing practices.

f) To establish and assert the rights of fishing communities to their customary territories in coastal zones under their national jurisdiction for fishing and habitation.

g) To promote a legal regime that will ensure the traditional and customary rights of fishing communities to the fishery under their national jurisdiction.

h) To promote the primary role of fisher peoples' organizations in managing fisheries and oceans, nationally and internationally.

i) To protect food security, both locally and worldwide, by sustaining fish stocks for the future and by preserving fish for human consumption.

j) To promote the equitable representation of fisher peoples' organizations in all relevant regional and international forums and to advocate for their recognition.

k) To play an active role in ensuring that states and transnational corporations comply with relevant international agreements and to oppose any trade agreements that threaten the livelihood of fishers.

l) To prevent the export of resource-collapse crises and of technologies and practices that lead to these crises.

m) To provide support for national and international struggles that are consistent with the objectives of the World Forum of Fisher Peoples (WFFP).

n) To encourage, assist, and support fisher peoples to organize themselves where they have not already done so.

o) To promote the right of fisher peoples to social security, safe working conditions, a fair income, and safety at sea, as well as their recognition as seafarers.

p) To improve the communication between fisher peoples and the scientific community through the exchange of knowledge and science.

q) To acknowledge and enhance the unique culture of fishing communities.

r) To restore our access to the rights and powers originally granted to us in the Charter of the United Nations.

MEMBERSHIP

Article 2: Active Members

a) All the organizations that share the objectives listed in Article 1 can become members of the World Forum of Fisher Peoples. However, under article 3, only one national organization per country can be a member of the Forum. This must be constituted democratically and can be a union, an association, a federation of cooperatives, or an autochthonous nation that depends on fishing for its livelihood. In any case, it must be representative of one or the other of the following groups:

- Fishermen, that is, all the people who are directly practicing fishing and are recognized in the different countries under the following titles:
    - People who practice fishing for subsistence;
    - Artisan fishermen;
    - Autochthonous or aboriginal communities that practice fishing;
    - Coastal and traditional continental fishermen;
    - Autonomous fishermen who practice small-scale fishing;
    - Crewmembers;
- Crewmembers who belong to groups not previously mentioned but who presently are part of the organizations described in the subparagraph (a) of the present article.
- Popular organizations rooted in fishermen's communities or including women engaged in the defense of fishing.
- Workers in the fishing sector whose activity consists in the transformation, selling (except for traders), and transporting of fish.
- Big companies and transnational societies, including their branches, that own fishing boats, do the catching, and transform and distribute fish products, and the companies that use destructive fishing methods or practice industrial aquaculture, cannot become members of the World Forum of Fisher Peoples.

Article 3: Admission of Participants

Only the Coordinating Committee can admit an active member. Exceptionally, the Coordinating Committee can, in full respect of the objectives indicated in article 2 of the present Statute, admit more than one organization per country, if these organizations represent an important part of the groups mentioned in subparagraphs 1 to 4 of article 2.

Article 4: Commitment

All the members of the World Forum of Fisher Peoples must formally accept the duties stipulated by and the content of the present Statute.

Article 5: Withdrawal

A participant can withdraw from the World Forum of Fisher Peoples by notifying the coordinating committee at least three months before the date of the Forum.

Article 6: Quotas

The deadlines and forms of payment are established by the coordinating committee. In very exceptional circumstances the Coordinating Committee can allow a member to postpone its payment for a specified period.

Article 7: Suspension of a Member

1. The Coordinating Committee can, after two calls, suspend a member that has not paid its quota. The suspension is annulled at the moment of the payment.
2. The Coordinating Committee can suspend a member whose actions go against the objectives of the Forum, provided that it has been heard before the decision is taken.
3. No suspended member can participate in the work of the Forum nor vote in the General Assembly.
4. It is possible to appeal any suspension to the General Assembly.

Article 8: Expulsion

The Coordinating Committee can expel a member if:

1. the corresponding Continental Council has recommended this measure;
2. the Coordinating Committee has given this member the opportunity to explain the actions at the origin of the expulsion.

STRUCTURES

Article 9: The Composition of the Forum

The World Forum of Fisher Peoples will consist of:

- General Assembly
- Coordinating Committee
- Five Continental Forums

THE GENERAL ASSEMBLY

Article 10: The General Assembly

The General Assembly is the highest body of the World Forum of Fisher Peoples. It is constituted by all the delegates of the member organizations. The Coordinating Committee can allow the representatives of organizations that are not members of the Forum to participate in the General Assembly as observers.

Article 11: Functions of the General Assembly

The General Assembly has the following functions:

- interpret and modify the Statute of the World Forum of Fisher Peoples;
- discuss questions of common interests and adopt resolutions on the issues brought to the agenda with a spirit of cooperation and free exchange of ideas;
- decide about the means needed to implement the decisions made and realize the objectives of the World Forum of Fisher Peoples;
- assign to the Coordinating Committee the tasks and missions considered necessary;
- organize elections in conformity with the dispositions of the present Statute;

- ratify, annul, or reflect upon the decisions of the Coordinating Committee and the Continental Councils.

Article 12: Representation of the Members at the General Assembly

1. Every country that has one least an active member has the right to be represented at the General Assembly by two delegates (a man and a woman).

2. Every active member has the right to nominate listeners or alternate delegates who can participate in the General Assembly with speaking rights.
3. The appointment of listeners or alternate delegates must be approved by the Coordinating Committee.
4. The exiting members of the Coordinating Committee automatically become delegates of the General Assembly when the new committee assumes its functions.

Article 13: Designation of the Delegates

1. The active members must communicate in writing to the Coordinating Committee the names of their delegates at least ninety days before the meeting of the General Assembly.
2. If a country has more than one member, they must try to agree on the choice of the delegates who will represent them. If they do not reach an agreement, they will be able to submit the problem to the Coordinating Committee, which will make a final decision on the matter.

Article 14: Decision-Making Process

The World Forum of Fisher Peoples will try to make all its decisions with the consensus of the participants.

Article 15: the Place of the General Assembly

Subject to the decisions made in the previous session, the General Assembly will take place at the site decided by the Coordinating Committee.

Article 16: Frequency of the General Assembly

The General Assembly takes place at least every three years.

THE COORDINATING COMMITTEE

Article 17: Composition of the Coordinating Committee

The Coordinating Committee will consist of two co-ordinators (a man and a woman), the general secretary, the treasurer, and the continental representatives, following the dispositions of article 18.

Article 18: Designation of the Continental Representatives

The Continental representatives in the Coordinating Committee will be chosen according to the following criterion: two representatives (a man and a woman) for each of these continents: Africa, America, Asia, Europe, Oceania.

Article 19: Functions of the Coordinating Committee

The Coordinating Committee is subject to the authority of the General Assembly and represents the World Forum of Fisher Peoples. Besides the tasks that the General Assembly assigns to it, it will have the following functions:

- promoting the creation of Continental Forums;
- planning and managing the activities of the World Forum of Fisher Peoples;
- organizing the General Assemblies of the members' representatives;
- drafting recommendations to be submitted to the General Assembly;
- implementing the decisions made by the General Assembly;
- representing the World Forum of Fisher Peoples to other organizations;
- admitting members in conformity with the present Statute;
- reporting its activities to the General Assembly;
- maintaining close ties among the members of the World Forum of Fisher Peoples between the sessions of the General Assembly;
- dealing with the preparation of the budget and the financial management of the World Forum of Fisher Peoples, according to the dispositions listed in article 29 of the present Statute;
- in general, taking all the necessary measures to achieve the objectives listed in the present Statute.

Article 20: Length of the Mandate of the Members of the Coordinating Committee

The mandate of the members of the Committee corresponds to the period between two General Assemblies and its length will be three years.

Article 21: Vacancy Inside the Coordinating Committee

1. If a member organization informs the Coordinating Committee that a member of the Committee no longer has a position justifying his/her mandate, the issue will be presented to the corresponding Continental Council. If the latter concludes that the information is correct, it will declare the position vacant.
2. If the evidence provided concerns a coordinator or the general secretary or the treasurer, the Coordinating Committee is authorized to adopt the necessary measures.
3. The Coordinating Committee is authorized, in case of a vacancy, to provide for the position of coordinator or general secretary or treasurer.
4. If a continental representative position, inside the Coordinating Committee, is vacant, the concerned Continental Council is authorized to provide for it.

Article 22: Procedure

The Coordinating Committee will establish its regulations and will decide its procedures and the frequency of its meetings. The Coordinating Committee will call a special meeting if two thirds of its members have presented a written request.

Article 23: Functions of the Coordinators

- Making all decisions jointly
- Convening the General Assembly and the Coordinating Committee.
- Presiding over the General Assembly and the Coordinating Committee.
- Coordinating the activity of the World Forum of Fisher Peoples in conformity with the objectives of the present Stature and the decisions made by the General Assembly and the Coordinating Committee.
- Facilitating the admission of new members to the World Forum of Fisher Peoples.

- Representing the World Forum of Fisher Peoples in other national and international forums or delegating other people to do it.
- In general, taking all the necessary measures to achieve the objectives listed in the present Statute and to have them approved by the Coordinating Committee.

Article 24: Functions of the Treasurer

- to take care of all the financial questions of the World Forum of Fisher Peoples;
- to act in conformity with the decisions of the Coordinating Committee;
- to make payments according to the decisions of the committee or the written instructions issued by jointly acting coordinators;
- to have the accounts of the World Forum of the Fisher Peoples approved by an external controller every year.

Article 25: Functions of the General Secretary

- to work under the direction of the coordinators;
- to guarantee the functioning of the office in charge of communication, implementation, and documentation;
- to preserve the official archives of the World Forum of Fisher Peoples;
- to complete all missions assigned to him of her by the Coordinating Committee.

Article 26: Terms of Office

The term of office of the coordinators, the treasurer, and the general secretary shall be three years, if the next General Assembly does not decide differently. Nobody will be elected for more than two consecutive terms, and this applies to all tasks.

CONTINENTAL FORUM

Article 27: Institution of the Continental Forum

1. The World Forum of Fisher Peoples recognizes five Continental Forums in conformity with the dispositions of article 18 of the present Statute.
2. Every Continental Forum is constituted by all the active mem-

bers of the corresponding continent mentioned in article 18.

3. The Continental Forums will freely decide their structure and the modalities of their functioning while respecting the objectives enunciated in article 1 and the other directives of the World Forum of Fisher Peoples. Every Statute will have to be approved by the Coordinating Committee.
4. The Continental Forums will have to identify their representatives for the Coordinating Committee.

Article 28: Continental Councils

1. Every Continental Forum will have to have a Continental Council, a continental coordinator, and other functionaries considered necessary.
2. The functions of the Continental Councils will be to ensure coordination and consultation among the member organizations in their respective continents, and to realize the program of the World Forum of Fisher Peoples.

GENERAL CONDITIONS

Article 29: Financing

a) The revenues of the World Forum of Fisher Peoples will come from:

- the dues paid by the members;
- donations or disbursements considered acceptable by the Coordinating Committee;
- any other source of revenue considered acceptable by the Coordinating Committee;

b) The Coordinating Committee, in fact, will not accept money from big corporations or other associations that act against the objectives of the World Forum of Fisher Peoples.

Article 30: Quorum

The quorum to propose meetings of the Forum will consist of the following:

1. General Assembly—two thirds of the delegates;
2. Coordinating Committee—more than 50 percent of the members of the Coordinating Committee.

Article 31: Dissolution

The World Forum of Fisher Peoples can be dissolved with the consensus of all those who are part of it.

Article 32: Offices

The location of the offices of the World Forum of Fisher Peoples will be identified from time to time by the Coordinating Committee.

Article 33: Official Languages

The official languages of the World Forum of Fisher Peoples will be English, French, and Spanish.

Article 34: Interpretation

The Coordinating Committee in the intervening period between General Assemblies will have the right to interpret the present Statute.

Article 35: The State

The World Forum of Fisher Peoples is an independent organization. Because of the authority conferred upon it by the General Assembly, the Coordinating Committee will be able to take any measure considered necessary to confer upon the Forum a juridical personality in conformity with the law in force in the country in which the Forum will conduct its activities.

APPENDIX THREE

WORLD FISHERIES DAY 2004[1]

On November 21, 2004, we are preparing once again to celebrate World Fisheries Day. This is a day dedicated to the fishermen of the world so they can realize that they are at the center of the development and management of marine resources. The World Forum of Fisher Peoples (WFFP) is glad to join these celebrations, in which all those who live thanks to the sea are participating: women and men who live by the sea and their families.

We are experiencing a crisis because the world's fish are overutilized by factory-like ships, by the scraping of the seabeds by bottom trawling, by excessive fishing and pollution. We must continue the struggle against the oversizing of boats, against excessive fishing, against destructive equipment of all types. The waters are polluted by domestic and industrial trash, by plastic materials and nuclear waste. If this continues, all the world's waters will become extremely polluted, and life on the planet will be in danger. We cannot destroy life on the planet in the name of development. The lives of millions of people depend on these waters. It is our responsibility and our duty to safeguard them for the good of humanity and the planet. Let us go forward together saying: let's protect the waters, let's protect life.

We want to conclude this message by citing the words of Chief Seattle, who in 1854 wrote: "Every part of this earth is sacred to my people. . . . Our dead never forget this beautiful earth, for it is the mother of the red man. We are part of the earth and it is part of us. . . . This shining water that moves in the streams and rivers is not just water but the blood of our ancestors. . . ; each ghostly reflection in the clear water of the lakes tells of events and memories in the life of my people. The water's murmur is the voice of my father's father. The rivers are our brothers; they quench our thirst. The rivers

[1] Sources www.wffp.org/indexcontent.asp?file1=worldfish.htm. Translated from French by Silvia Federici.

carry our canoes and feed our children. . . . [T]he earth does not belong to man, man belongs to the earth. . . . If men spit upon the ground, they spit upon themselves. . . . All things are connected and depend on each other."

This is how the world fishermen join to promote and protect our mother earth, with all its natural resources, the soil and the forest. We celebrate World Fisheries Day on November 21, affirming, "We will take care of you, our mother ocean, and the marine resources, until [for as long as] we have a breath of life."

BIBLIOGRAPHY

Administration of the Province of Venice (Amministrazione della provincia di Venezia), ed. *La pesca nella laguna di Venezia.* Venezia: Albrizzi, 1981.

Aguiton, C. *Il mondo ci appartiene. I nuovi movimenti sociali.* Milano: Feltrinelli, 2001.

Aguiton, C., et al. *Globalizzazione delle resistenze e delle lotte. L'altraDavos*, Bologna: Emi, Editrice missionaria italiana, 2000.

Altobrando, A., and Turus, G., eds. *Biodiversità.* Padova: Esedra, 2005.

Antonini, L., Marcato, A., Rallo, G. *Gli antichi mestieri delle valli.* Mestre: Museo del Territorio delle Valli e Laguna di Venezia, 2002.

Associazione Culturale "El Fughero," ed. *La pesca in mare. Metodi, tecniche, esperienze di vita.* Venezia: Salvano, 1989.

Baldi, G., et al. *Dal Testo alla Storia. Dalla storia al testo*, vol. II/1. Torino: Paravia, 1995.

Bales, K. *Disposable People: New Slavery in the Global Economy.* Berkeley: University of California Press, 1999.

Baudelaire, C. *I fiori del male.* Milano: Feltrinelli, 2003.

Bolzoni, A. "*Fuga dal Mediterraneo, i tonni sono scomparsi,*" in *La Repubblica*, 9 maggio 2003.

Bonefeld, W., ed. *Subverting the Present, Imagining the Future. Insurrection, Movement, Commons.* Brooklyn, NY: Autonomedia, 2008.

Bono, S. *Corsari nel Mediterraneo.* Milano: Mondadori, 1993.

Braudel, F. *The Mediterranean and the World in the Age of Philip II*, Vols. 1 and 2. New York: Harper & Row, 1972.

———. *Memorie del Mediterraneo.* Milano: Bompiani, 1998.

Caffarena, A. *Governare le onde. Le prospettive della cooperazione internazionale.* Milano: FrancoAngeli, 1998.

Canepa, E. *Impatto dei sistemi "antifouling," utilizzati per le imbarcazioni da pesca, sull'ecosistema"* (Impact on the ecosystem of the 'antifouling' systems used for fishing boats) in *Gestione delle risorse biologiche e sviluppo sostenibile: le attività di pesca nella Riviera di Levante.* Genova: LII Pubblicazione della Sezione di Scienze Geografiche dell'Università di Genova, Brigatti, 1998.

Cannavò, S. *Porto Alegre. Capitale dei movimenti.* Roma: Manifestolibri, 2002.

Capitalismo Natura Socialismo, n. 1, gennaio–aprile, 1995.

Carbone, M. "Le milieu marin et le développement durable," in *Le Courrier ACP-UE*, n. 193, juillet–août 2002.

Carra, L., and Terragni, F. *Il futuro del cibo. Gli alimenti transgenici.* Milano:

Garzanti, 1999.

Centro nuovo modello di sviluppo. *Nord Sud. Predatori, predati e opportunisti*. Bologna: Emi, Editrice missionaria italiana, 1993

Ceri, P. *Movimenti globali. La protesta nel XXI secolo*. Roma-Bari: Editori Laterza, 2002.

Chossudovsky, M. *The Globalization of Poverty: Impacts of IMF and World Bank Reforms*. London, Zed Books Ltd., 1999.

Ciancetti, E., and Jirillo, R. *Emergenza Ambiente*. Roma: Città Nuova, 1991.

Ciavorella, G. *Mito Poesia e Storia*. Torino: Il Capitello, 1990.

Common Sense n. 23, 1998

The Commoner n. 6, 2002, in www.thecommoner.org.

The Commoner n. 8, 2003, in www.thecommoner.org.

The Commoner n. 12, 2007, in www.thecommoner.org.

Coppo, L. *Terra gamberi contadini ed eroi*. Bologna: Emi, Editrice missionaria italiana, 2002.

Corbin, A. *L'invenzione del mare*. Venezia: Marsilio, 1990.

Le Courrier ACP-UE, n. 191, mars–avril 2002.

Le Courrier ACP-UE, n. 192, mai–juin 2002.

Le Courrier ACP-UE, n. 193, juillet–août 2002.

Dalla Costa, M., "Capitalism and reproduction" in Werner Bonefeld ed., *Subverting the Present, Imagining the Future. Insurrection, Movement, Commons*. Brooklyn, NY: Autonomedia, 2008, and in *The Commoner* n. 8, 2003, in www.thecommoner.org (Translated from the Italian "Capitalismo e riproduzione," in *Capitalismo Natura Socialismo*, n. 1, gennaio–aprile 1995).

———. "Some Notes on Neoliberalism, on Land and on the Food Question," in *Canadian Women Studies, Les Cahiers de la femme*. New York, Ontario, Canada (Spring 1997). (Translated from the Italian "Neoliberismo, terra e questione alimentare," in *Ecologia politica–CNS*, n. 1, febbraio 1997.)

———. "The Native in Us, the Land We Belong to" in *Common Sense* n. 23, 1998, and in *The Commoner*, n. 6, 2002, in www.thecommoner.org (Translated from the Italian, "L'indigeno che è in noi, la terra a cui apparteniamo," in A. Marucci, a cura di, *Camminare domandando. La rivoluzione zapatista*. Roma: DeriveApprodi, 1999).

———. "Perché i pesci saltino nell'orto. Biodiversità e salute nei movimenti per un'agricoltura contadina e una pesca artigianale," in Altobrando, A., Turus, G., a cura di, *Biodiversità*. Padova: Esedra editrice, 2005.

———. "Food as Common and Community," in *The Commoner*, n. 12, 2007, in www.thecommoner.org.

Dalla Costa, M., and Dalla Costa, G. F., eds. *Women, Development and Labor of Reproduction*. Trenton, NJ, and Asmara, Eritrea: Africa World

Press, 1999. (Translated from the Italian, *Donne, sviluppo e lavoro di riproduzione. Questioni delle lotte e dei movimenti*. Milano: FrancoAngeli, 1996.)

Della Porta, D. *I new global*. Bologna: Il Mulino, 2003.

Della Seta, R. *La difesa dell'ambiente in Italia. Storia e cultura del movimento ecologista*. Milano: FrancoAngeli, 2000.

De Marcellus, O. "Divisions and missed opportunities," in *Metamute*, n. 28, Summer–Autumn 2004.

De Nardis, F. *Cittadini globali*. Roma: Carocci, 2003.

Ecologia politica–CNS, n. 1, febbraio 1997.

Ecologia politica–CNS, n. 3–4, agosto–dicembre 2003.

Ecologist Asia, vol. 3, July–August 1995.

Economia e ambiente, n. 1–2, gennaio–aprile 2001.

Eldredge, N. *Life in the Balance: Humanity and the Biodiversity Crisis.* Princeton: Princeton University Press, 1998.

Ferrari, F. "I pescatori dell'Adriatico dalle lagune alle grandi migrazioni," in *Chioggia, rivista di studi e ricerche*, n. 12, pp. 129–137, Chioggia, 1988.

———. "Alle origini della biologia marina in Italia. Una lettura antropologica delle origini di una scienza," in *Notiziario Sibm*, n. 39, pp. 44–46, Genova, aprile 2001.

———. "Il futuro della pesca in Alto Adriatico," in *Notiziario Sibm*, n. 45 pp. 63–68, Genova, maggio 2004.

Fishermen's Voice, December 2009, available at: http//:www.fishermenvoice.com.

Fontana, C., and Giacci, M. *Gli alberi e la foresta*, vol. A. Padova: Cedam, 2001.

Forti, M. *La signora di Narmada. Le lotte degli sfollati ambientali nel Sud del mondo*. Milano: Feltrinelli, 2004.

Francescato, G., et al. *No global. Da Seattle a Porto Alegre*. Milano: Scheiwiller, 2002.

Frey, B. S. *Economia politica internazionale*. Milano: FrancoAngeli, 1987.

Gallino, L. *Globalizzazione e disuguaglianze*. Roma-Bari: Editori Laterza, 2002.

George, S. *A Fate Worse than Debt*. London: Penguin Books, 1988

———. *Fermiamo il Wto*. Milano: Feltrinelli, 2002.

Garaguso, G. C., and Marchisio, S., a cura di, *Rio 1992: Vertice per la Terra*. Milano: FrancoAngeli, 1993.

Giansanti, P. "Afgani, otto anni. Valore: 15 chili di gamberetti," in *Sette*, supplemento a *Il Corriere della sera*, 29 novembre 2002.

Held, D., and McGrew, A. *Globalization/Anti-Globalization*. Cambridge, UK: Polity Press, 2002.

Hobbes, T. *Leviathan*. New York: World Publishing Company, 1963.

Hugo, V. *L'uomo che ride*. Milano: Garzanti, 1988.

International Organization, n. 3, 1982.

Isaacs, J. D., "Forme di vita nell'oceano," in *Le Scienze*, n. 16, dicembre 1969.

Janni, P. *Il mare degli Antichi*. Bari: Edizioni Dedalo, 1996.

Jonas, H. *Il principio responsabilità. Un'etica per la civiltà tecnologica*. Torino: Einaudi, 1990.

Lasserre, P. "Coastal Lagoons. Sanctuary Ecosystem, Cradles of Culture, Targets for Economic Growth," in *Nature and Resources*, vol. 15, 1979.

Laureti, L. *Economia e sviluppo della pesca. Lo sviluppo sostenibile*. Padova: Cedam, 2000.

Lessing, E. a cura di, *L'Odissea. L'epopea omerica nel racconto fotografico di E. Lessing*. Alba: Edizioni Paoline, 1989.

"Limes," *I popoli di Seattle*, n. 3, 2001.

Madhusoodanan, G. "Il modello Kerala alla prova dell'ambientalismo," in *Ecologia politica–CNS*, n. 3–4, agosto–dicembre 2003.

Mainardi, D. *L'animale irrazionale. L'uomo, la natura e i limiti della ragione*. Milano: Mondadori, 2001.

Mander, G., and Goldsmith, E. *The Case against the Global Economy, and for a Turn toward the Local*. San Francisco: Sierra Club Books, 1996.

Marchesini, R. *La fabbrica delle chimere*. Torino: Bollati Boringhieri, 1999.

Marucci, A., a cura di. *Camminare domandando. La rivoluzione zapatista*. Roma: DeriveApprodi, 1999.

Massard-Guilbaud, G. "De la 'part du milieu' *à l'histoire de l'environnement*," in *Le mouvement social*, n. 200, juillet–septembre 2002.

Mela, A., et al., *Sociologia dell'ambiente*. Roma: Carocci, 1998.

Merchant, C. *The Death of Nature. Women, Ecology and the Scientific Revolution*. New York: Harper and Row, 1980.

Mercurio, R. "Una rete di protezione per l'ambiente Mediterraneo," in *Economia e ambiente*, n. 1–2, gennaio–aprile, 2001.

Metamute, n. 28, Summer–Autumn 2004.

Michelet, J. *Il mare*. Genova: Il Melangolo, 1992.

Midnight Notes. *The New Enclosures*. Brooklyn, NY: Autonomedia, 1990.

Mollat du Jourdin, M. *L'Europa e il mare*. Bari: Laterza, 1993.

Montale, E. *Ossi di seppia*. Milano: Mondadori, 1991.

Montanari, M. *L'Europa a tavola. Storia dell'alimentazione dal Medioevo a oggi*. Bari: Laterza, 1999.

Morante, E. *L'isola di Arturo*. Torino: Einaudi, 1957.

More, T. *The 'Utopia' and the History of Edward V*, ed. Maurice Adams, ed. London: Walter Scott, 1980.

Morrone, A. *L'altra faccia di Gaia. Salute, migrazione e ambiente tra Nord e Sud del pianeta*. Roma: Armando, 1999.

Moussis, N. *Europa protagonista*. Milano: Etas libri, 1991.

"Le mouvement social," n. 200, juillet–septembre 2002.

"Il Mulino. Rivista bimestrale di cultura e politica," n. 399, gennaio 2002.

N. Myers, ed. *Gaia Atlas of Future Worlds: Challenge and Opportunity in an Age of Change*. London: Robertson McCarta Ltd., 1990.

Nature and Resources, vol. 15, 1979.

Neveu, E. *I movimenti sociali*. Bologna: Il Mulino, 2001.

Onida, F. "La globalizzazione aumenta o riduce disuguaglianze e povertà?" in *Il Mulino. Rivista bimestrale di cultura e politica*, n. 399, 1/2002.

Papisca, A., and Mascia, M. *Le relazioni internazionali nell'era dell'interdipendenza e dei diritti umani*, 3a ed. Padova: Cedam, 2004.

Penzo, P., a cura di, *Le attività ittiche a Chioggia: le realtà di oggi, le prospettive e gli interventi necessari*. Venezia: Libreria editrice Il Leggio, 1997.

Pontara, G., a cura di, *Teoria e pratica della non violenza*. Torino: Einaudi, 1973.

———. *Guerra, disobbedienza civile, non violenza*. Torino: Edizioni Gruppo Abele, 1996.

———. *La personalità non violenta*. Torino: Edizioni Gruppo Abele, 1996.

———. *Breviario per un'etica quotidiana*. Parma: Edizioni Pratiche, 1998.

Popper, K. *In Search of a Better World*. New York: Psychology Press, 1996.

Rees, J. *Natural Resources. Allocation, Economics and Policy*. London: Routledge, 1990.

La Repubblica, 9 maggio 2003.

Ricoveri, G. *Beni comuni tra tradizione e futuro*. Bologna: Emi, Editrice missionaria italiana, 2005.

Rienstra, D. "*La pêche au thon: le secteur primordial*, " in *Le Courrier ACP-UE*, n. 192, mai–juin 2002.

Ruesch, H. *Imperatrice nuda*. Milano: Rizzoli, 1976.

Sachs, W. *Global Ecology*. London: Zed Books, 1993.

———. ed. *The Development Dictionary*. Johannesburg: Wits University Press, 1993.

La Sacra Bibbia. Roma: Edizione ufficiale della Cei, Ueci, 1986.

Schmitt, C. *Terra e mare*. Milano: Giuffré, 1986.

———. *Il nomos della terra*. Milano: Adelphi, 1991.

Le Scienze, n. 16, dicembre 1969.

Sepulveda, L. *Il mondo alla fine del mondo*. Parma: Guanda, 1994.

Serra, C. *Le biotecnologie*. Roma: Editori Riuniti, 1998.

Serra, V., a cura di, *Le parole del mare. Libero viaggio nell'oceano letterario*.

Milano: Baldini & Castoldi, 2002.

Sette, supplemento a "Il Corriere della sera," 29 novembre 2002.

Shiva, V. *Staying Alive: Women, Ecology and Survival in India.* New Delhi: Kali for Women and London: Zed Books, 1988 (also available as *Staying Alive: Women, Ecology and Development.* Cambridge, MA: South End Press, 2010).

———. *Monocultures of the Mind: Biodiversity, Biotechnology and Agriculture.* New Delhi: Zed Press, 1993.

———. *Biopiracy: The Plunder of Nature and Knowledge*. Cambridge, MA: South End Press, 1997.

———. *Stolen Harvest: The Hijacking of the Global Food Supply*. Cambridge, MA: South End Press, 1999.

———. *Patents, Myths and Reality*. India: Penguin, 2001.

———. *Water Wars; Privatization, Pollution, and Profit.* Cambridge, MA: South End Press, 2002.

Thomas, K. *Man and the Natural World: Changing Attitudes in England, 1500–1800*. London: Penguin Press,1983.

Vallenga, A. *Ecumene oceano*. Milano: Mursia, 1995.

Verne, J. *Twenty Thousand Leagues Under the Sea*. Blackburg, Virginia: Wilder Publications, 2008.

Wallach, L., and Sforza, M. *WTO*. Milano: Feltrinelli, 2001.

Wijkman, P. M. "International Organization," in *Managing the Global Commons*, n. 3, 1982.

Zamperetti, F., and Dalla Costa, G. F. *Sharing Trust, Women and Microcredit in Eritrea.* Trenton, NJ: Africa World Press, 2008 (Translated from the Italian, *Microcredito donne e sviluppo. Il caso dell'Eritrea.* Padova: Cleup Editrice, 2003).

About the Authors

MARIAROSA DALLA COSTA is an influential author and international feminist who has devoted her theoretical and practical efforts to the study of the female condition in capitalist development. Starting with Potere Operaio, Lotta Femminista, and the International Wages for Housework Campaign, Dalla Costa has for decades been a central figure in the development of autonomist thought in a wide range of anticapitalist movements. Her seminal book *The Power of Women and the Subversion of the Community*, coauthored with Selma James (Falling Wall Press, 1972), has been translated into seven languages. Her writings, reflecting on the role of social reproduction in the organization of autonomy as well as the historic development of capital, have consistently been staged within and through social struggles and movements organizing around the questions of land, agriculture, food, and the commons.

Dalla Costa's writings have been published in English, Spanish, Portuguese, French, German, Turkish, Korean, Greek, and Japanese. Many of her articles are available in *The Commoner* (www.commoner.org.uk). In 2009, an anthology of her writings, *Dinero Perlas y Flores en la Reproduccion Feminista*, was published in Spain by Akal, Madrid. Her English publications include *Women, Development and Labor of Reproduction* (Africa World Press, 1999), coedited with G. F. Dalla Costa, *Gynocide, Hysterectomy, Capitalist Patriarchy and the Medical Abuse of Women* (Autonomedia, 2007), *Our Mother Ocean: Enclosures, Commons, and the Global Fishermen's Movement* (Common Notions, 2014), as well as *Family, Welfare, and the State: Between Progressivism and the New Deal* (Common Notions 2015).

Mariarosa Dalla Costa previously taught at the Department of Political and Juridical Science and International Studies at the University of Padua, Italy.

MONICA CHILESE is a researcher and political sociologist. She worked in the Faculty of Political Science at the University of Padua, where she devoted her studies to questions of ecology, giving special attention to the marine environment, the impoverishment of fisheries, and corresponding social problems. Today she does social research at a scholarly institute in northeast Italy.

About the Translator

SILVIA FEDERICI is a feminist writer, teacher, and activist. Her most recent book is *Revolution at Point Zero: Housework, Reproduction, and Feminist Struggle* (Common Notions/PM Press, 2012). Along with Dalla Costa, she cofounded the International Feminist Collective in 1972, which launched the Wages for Housework campaign internationally. In the 1990s, after a period of teaching and research in Nigeria, she was active in the antiglobalization movement and the U.S. anti–death penalty movement. She is one of the cofounders of the Committee for Academic Freedom in Africa, an organization dedicated to generating support for the struggles of students and teachers in Africa against the structural adjustment of African economies and education systems. From 1987 to 2005 she also taught international studies, women's studies, and political philosophy at Hofstra University in Hempstead, NY.

Federici's decades of research and political organizing include a long list of publications on philosophy and feminist theory, women's history, education, culture, international politics, and more recently, the worldwide struggle against capitalist globalization and for a feminist reconstruction of the commons. Her steadfast commitment to these issues resounds in her focus on autonomy in the building of what she calls self-reproducing movements as a powerful challenge to capitalism through the construction of new social relations.

COMMON NOTIONS is a publishing house and programming machine that fosters the collective formulation of new directions for living autonomy in everyday life.

We aim to translate, produce, and circulate tools of knowledge production utilized in movement-building practices. Through a variety of media, we seek to generalize common notions about the creation of other worlds beyond state and capital. Our publications trace a constellation of historical, critical, and visionary meditations on the organization of both domination and its refusal. Inspired by various traditions of autonomism—in the U.S. and internationally, historically and emerging from contemporary movements—we aim to provide tools of militant research in our collective reading of struggles past, present, and to come. Common Notions regularly collaborates with editorial houses, political collectives, militant authors, and visionary designers around the world.

Our political and aesthetic interventions are dreamt and supported in collaboration with Antumbra Designs.

www.commonnotions.org
info@commonnotions.org